MASTER MATHEMATICS TEACHERS

This practical guide invites teachers to take a journey towards masterly mathematics teaching using the experiences and lessons learnt across five Chinese provinces, Anhui, Beijing, Jiangsu, Jiangxi and Tianjin.

On this journey, you will gain a thorough understanding of: (1) the quality and characteristics of master mathematics teachers' teaching, (2) the quality of mathematics learning they have nurtured amongst their students in affective, metacognitive and cognitive dimensions and (3) the teaching-learning mechanisms that underpin excellent performance in the three dimensions. Alongside the quantitative and qualitative evidence on educational excellence, you will also delve deep into the trajectories and processes of professional development that generate professional excellence amongst master mathematics teachers and their peers within and across primary schools, up and down China.

Building on data collected with 70 master mathematics teachers and 3,178 students and from teaching research events at the school, municipal, provincial and national levels, the Master Mathematics Teachers (MasterMT) project is, to date, the first that has looked at the legendary tribe of master mathematics teachers in China at such a large scale, and with such breadth and depth. The book offers invaluable insights for any teacher or teacher educator who wants to improve mathematics teaching and learning and generate educational and professional excellence in primary schools and beyond.

More information on the Master Mathematics Teachers (MasterMT) project can be accessed at https://mastermt.org/.

Zhenzhen Miao is Assistant Professor in Mathematics Education at Jiangxi Normal University, China. Zhenzhen teaches both initial teacher education (ITE) and research courses in Mathematics Education to both undergraduates and postgraduates and is an active contributor to teacher professional development (PD) at local schools, provincial teaching research conferences and the National PD Programme for Rural Teachers in China. Her research focuses on mathematics learning, teaching, ITE and teacher PD in both local and global contexts. She has a PhD from the University of Southampton, UK.

Christian Bokhove is Professor in Mathematics Education at Southampton Education School, University of Southampton, UK. He is also Director of Research at the School and has co-led their Mathematics, Science and Health Education research centre. He has a PhD from Utrecht University, Netherlands.

David Reynolds, CBE, was Deputy Pro Vice Chancellor, Head of Department and Professor of Education at Swansea University, UK. Since 2021, he has been Distinguished Professor of Education at Hangzhou Normal University, China. He has published extensively in the areas of educational effectiveness, comparative education and educational policy (over 20 books, 150 articles) and has lectured, consulted and/or researched in over 70 countries.

'Chinese students' mathematics performance is consistently good in international comparative studies. The role and effort of good mathematics teachers is indispensable. What is the secret of a good mathematics teacher in China? This book will tell you all about it. The MasterMT project aims to unfold the secret behind best teaching practice in primary mathematics in China via a mixed methods study. The study put focuses on three dimensions of mathematics teaching, namely action, mind and pattern. The findings delineate vivid pictures of master mathematics teachers' individual and collective effects of various factors at multiple levels of student achievement in mathematics knowledge, skills and thinking; their perceptions of schooling and math learning; their professional trajectories. Educators, researchers, teachers and parents will enjoy this book.'

Mok Ah Chee Ida, *Associate Professor, Faculty of Education,*
The University of Hong Kong

'Education policy across the world is often based on claims made about practices in other countries, and in Mathematics China has been one of the countries often cited. However, much of this commentary is based on overinterpreting international studies such as PISA, with very little research on what mathematics teaching in China is really like. This important book reports on a major study of Master Mathematics Teachers in five Chinese provinces, and provides us with rigorous data and analysis on their teaching practice. This is linked to student outcomes, not just in the cognitive sphere, but in the affective and metacognitive spheres as well. All of these make this book a must-read for anyone interested in mathematics education, and how we can improve it.'

Daniel Muijs, *Professor of Education and Head of the School of Social Sciences,*
Education and Social Work, Queen's University Belfast

MASTER MATHEMATICS TEACHERS

Educational and Professional Excellence in Chinese Primary Schools

Zhenzhen Miao, Christian Bokhove and David Reynolds

Routledge
Taylor & Francis Group

LONDON AND NEW YORK

Designed cover image: © Getty Images

First published 2024
by Routledge
4 Park Square, Milton Park, Abingdon, Oxon OX14 4RN

and by Routledge
605 Third Avenue, New York, NY 10158

Routledge is an imprint of the Taylor & Francis Group, an informa business

© 2024 Zhenzhen Miao, Christian Bokhove and David Reynolds

The right of Zhenzhen Miao, Christian Bokhove and David Reynolds to
be identified as authors of this work has been asserted in accordance with
sections 77 and 78 of the Copyright, Designs and Patents Act 1988.

British Library Cataloguing-in-Publication Data
A catalogue record for this book is available from the British Library

ISBN: 978-0-367-65125-1 (hbk)
ISBN: 978-0-367-65123-7 (pbk)
ISBN: 978-1-003-12792-5 (ebk)

DOI: 10.4324/9781003127925

Typeset in Times New Roman
by Apex CoVantage, LLC

*We dedicate this book to
all the students and master mathematics teachers
who open their classrooms to us.*

CONTENTS

TABLES

FIGURES

ACRONYMS

AF	Affective Outcomes
AIC	Akaike's Information Criterion
AIM	Achievement in Mathematics
ANOVA	Analysis of Variance
ATM	Attitudes towards Mathematics
BIC	Bayesian Information Criterion
CCSSM	Common Core State Standards: Mathematics
CCSSO	Council of Chief State School Officers
CFA	Confirmatory Factor Analysis
CFI	Comparative Fit Index
CSMS	Concepts in Secondary Mathematics and Science
DfE	Department for Education
EEF	Education Endowment Foundation
EFA	Explorative Factor Analysis
EGM	Perceived Teaching Engagement
EMT	Effective Mathematics Teaching
ES	Effect Size
GDP	Gross Domestic Product
HCF	Highest Common Factor
ICC	Intraclass Correlation
ICCAMS	Improving Competence and Confidence in Algebra and Multiplicative Structures
ICT	Information and Communication Technologies
IEA	International Association for the Evaluation of Educational Achievement
IF	ISTOF

ISTOF	International System for Teacher Observation and Feedback
IT	Information Technology
ITE	Initial Teacher Education
KC	Knowledge of Cognition
KS	Key Stage
LCM	Lowest Common Multiple
LV	Latent Variable
MasterMT	Master Mathematics Teachers
MC	Metacognition
MLM	Multilevel Modelling
MNSQ	Mean Squares
MoE	Ministry of Education
MQ	Mathematical Quality (MQI whole-lesson code)
MQI	Mathematical Quality of Instruction
MQIF	MQI and ISTOF
MSEM	Multilevel Structural Equation Modelling
MTS	Master Teacher Studio
NGA	National Governors Association
NMAP	National Mathematics Advisory Panel
OECD	The Organisation for Economic Co-operation and Development
OTL	Opportunity to Learn
PCK	Pedagogical Content Knowledge
PD	Professional Development
PFC	Prefrontal Cortex
PI	Principal Investigator
PISA	Progress for International Student Assessment
PLC	Professional Learning Community
QUAL	Qualitative
QUAN	Quantitative
RC	Regulation of Cognition
RCT	Randomised Controlled Trial
RMSEA	Root Mean Square Error of Approximation
RP	Ratio and Proportion
SBL	Sense of School Belonging
SD	Standard Deviation
SEM	Structural Equation Modelling
SES	Socioeconomic Status
SMK	Subject Matter Knowledge
SRMR	standardised root mean square residual
SRMRb	SRMR between
SRMRw	SRMR within
STEM	Science, Technology, Engineering and Mathematics
TALIS	Teaching and Learning International Survey

TAM	Test Analysis Modules (an R package)
TIMSS	Trends in International Mathematics and Science Study
TL	Teaching and Learning
TLI	Tucker-Lewis index
TRG	Teaching Research Group

PREFACE

The book is based on the Master Mathematics Teachers (MasterMT) project with classroom data collected in 2018 through 2019 in five Chinese provinces – Anhui, Beijing, Jiangsu, Jiangxi and Tianjin. The MasterMT project focuses on the historically understudied cohort of legendary practitioners, the master mathematics teachers in China.

Synthesising existing research, our previous study on effectiveness of mathematics teaching across two countries and Zhenzhen's updated knowledge in working with student teachers, teachers and teaching research officials in China, we realise that the final inch into quality mathematics teaching is to study the best practitioners, that is, master teachers, who have been pioneering the teaching profession and playing a leading role in their peers' PD from bottom up in China for decades.

Existing empirical studies about them are primarily qualitative and of small scale, with an explicit yet somewhat narrow focus on a singular phenomenon, such as the development of a particular exemplary lesson or on teacher beliefs about PD through small-scale interviewing or survey. Building on existing research, we further push the boundary of research on master teachers from the perspective of cases to cohorts, from singular fragmented aspects to the wide-angle view of comprehensive aspects.

Existing empirical studies on mathematics teaching show a methodological divide or – put it another way – paradigmatic divide, with some applying quantitative analyses to often 80 to 100 'stand-alone' lessons per nation (e.g., the TIMSS Video Studies and the most recently TALIS Video Study: Hiebert et al., 2003; OECD, 2020; Stigler & Hiebert, 1999) and others applying qualitative interpretation of characteristics of stand-alone or consecutive lessons (e.g., the Learner's Perspective Study: Clarke et al., 2006).

Existing empirical studies on mathematics learning show a conceptual disparity in the kinds of learning that should be studied. The recurring international assessments have been focusing on cognitive outcomes, with affective variables collected but rarely captured by the spotlight. There are studies asserting the important role of affective outcomes but risking falling in the classical *either-or* trap hidden beneath almost all facets of educational research/practice and lacking attention to alternative competences such as metacognition which is the crucial intrinsic component for learning strategically.

The MasterMT project is the first that studied a large sample of master mathematics teachers in China, looked at teaching and learning in their classrooms in multiple dimensions and scrutinised both processes and outcomes of maths teacher PD, integrating both quantitative and qualitative methods and data. Learning-wise, we have measured three types of learning outcomes: affective learning outcomes, cognitive learning outcomes and metacognitive learning outcomes, giving particular attention to the cultivation of the three in our interpretation of the classroom-process data. Teaching-wise, we have looked at both *stand-alone lessons*, one lesson per teacher, by all teacher participants and *consecutive lessons* by a master teacher over 40 school days. This gives the project's findings, in terms of just teaching, invaluable breadth (QUAN) and depth (QUAL). Pulling together learning and teaching, we gained insights into the processes, outcomes and mechanisms of teaching and learning in master teachers' classes. Equally importantly, we have listened to master teachers about their views on mathematics teaching and learning, their comments on their own and colleagues' lessons and their reflection upon the paths towards masterly teaching over decades. In addition, we have tried to step into their shoes and immersed ourselves in mathematics teaching research meetings/conferences at the school, municipal, provincial and national levels, gathering first-hand information on maths teachers' PD *in action*.

In this book, we take the journey towards the unpacking of the masterly teaching practice and teacher professional development in primary mathematics based on the following data: (1) 109 lessons delivered by 70 master maths teachers to 3,178 students and analysed with a hybrid of observation systems; (2) teacher demographics, professional status, maths teaching beliefs and self-efficacy collected with a teacher questionnaire; (3) post-lesson interviews with the teachers; (4) professional development trajectories reported by master teachers in writing; (5) students' metacognitive learning outcomes; (6) students' cognitive learning outcomes; (7) students' affective learning outcomes; (8) case study data about a master maths teacher's work spanning 40 school days (23 April–25 October 2018), including 40 video-recorded lessons, post-lesson interviews, school-based teaching research group meetings and other data collected during the longitudinal fieldwork; (9) our participatory observation of teaching research conferences at the municipal, provincial and national levels.

Through the analyses of quantitative and qualitative data and systematic integration of both, this book will (1) update our knowledge as to what makes masterly mathematics teaching with a rich collection of models and explanations, (2) inform practitioners with concrete evidence and rich techniques from their pioneering peers for improving teaching, and (3) contribute to research and practice with latest, rich and robust evidence from the extraordinary 'tribe' of master mathematics teachers.

Zhenzhen Miao
Christian Bokhove
David Reynolds
December 2022

References

Clarke, D., Keitel, C., & Shimizu, Y. (Eds.). (2006). *Mathematics classrooms in twelve countries: The insider's perspective*. Rotterdam: Sense Publishers.

Hiebert, J., Gallimore, R., Garnier, H., Givvin, K. B., Hollingsworth, H., Jacobs, J., . . . Stigler, J. (2003). *Teaching mathematics in seven countries: Results from the TIMSS 1999 Video Study*. Washington, DC: National Center for Educational Statistics.

OECD. (2020). *Global teaching insights: A video study of teaching*. OECD Publishing. https://doi.org/10.1787/20d6f36b-en.

Stigler, J. W., & Hiebert, J. (1999). *The teaching gap: Best ideas from the world's teachers for improving education in the classroom*. New York: The Free Press.

ACKNOWLEDGEMENTS

This project and the book would not be possible without the help and support from a number of people in educational practice, research and publishing sectors and from our own life. We owe a huge debt of gratitude to all of them.

We would like to thank all the students and the master teachers for taking part in the project. Our enormous gratitude goes to the following teaching research officials (TROs, in the alphabetic order of provinces): Mr Tao Hu (TRO in Anhui); Ms Cunli Fan, Mr Kechen Liu and Ms Yanwei Wang (TROs in Beijing); Mr Qing-song Guo and Mr Qihua Zhang (TROs in Jiangsu); Mr Xianqing Song (TRO in Jiangxi); Ms Zhanjie Ren (TRO in Tianjin), for their kind support in helping us recruit master teachers. We also owe huge gratitude to Mr Yongchun Wang, Chief-Editor for Primary Mathematics at the People's Education Press, for introducing TROs in Beijing to us. We are immensely grateful to a number of headteachers and heads of the mathematics department from the participating schools for their kind support during the data collection, though for confidentiality we cannot name them here. We would also like to thank Jiangxi Normal University for funding the project and the three undergraduate students, Ms Jinying Fu, Ms Yuwen Zhang and Ms Yuanyuan Zhu, for their contribution to the grading of the test. Huge grati-tude also goes to the following people and organisation (in the alphabetical order of surnames or acronyms): Professor Heather Hill at Harvard Graduate School of Education for permitting us to use the MQI observation system on the project; Professor Jeremy Hodgen at Institute of Education, University College London for permitting us to reuse the test items developed/used on the CSMS/ICCAMS projects; the International Association for Educational Evaluation and Assessment (IEA) for permitting us to use the background items and several scales (as cited in the book) of the TIMSS 2015 Student Questionnaire (Grade 4) on the project; Pro-fessor Ina V. S. Mullis at the TIMSS & PIRLS International Study Center, Lynch

School of Education, Boston College for kindly responding to and forwarding our email about reuse permission to the IEA. We are also deeply grateful to our Editor at Routledge, Ms Katie Peace, for her genuine interest in the project and the book from the very beginning, all the patience, professional support and trust that we have received from her throughout the process.

With huge gratitude towards all who have kindly offered us help and support, we take entire responsibility for the quality and findings of the project.

Zhenzhen Miao
Christian Bokhove
David Reynolds

Personal acknowledgements to family

I would like to thank my beloved husband who has been effortlessly supporting me throughout the project and my son who has always been my greatest inspiration. I would never have been able to pursue my dream in academia without their love, support, encouragement, friendship and patience over the past 12 years since I started my study in Southampton.

Zhenzhen Miao
December 2022

AUTHORS

Zhenzhen Miao is Assistant Professor in Mathematics Education at Jiangxi Normal University, China. She conducted her PhD study (2016) on the effectiveness of mathematics teaching in primary schools in China and England at the University of Southampton, UK. Upon completing her PhD, she returned to China and took up an Assistant Professor position at Jiangxi Normal University in 2017, under the University's Outstanding Returnee Scholar scheme. The current project was funded by a grant (CNY 100,000) that came with the scheme. Zhenzhen teaches both initial teacher education (ITE) and research courses in Mathematics Education to both undergraduates and postgraduates and is an active contributor to teacher professional development (PD) at local schools, provincial teaching research conferences and the National PD Programme for Rural Teachers in China. Her research focuses on mathematics learning and teaching as well as mathematics ITE and teacher PD in both the local and global contexts.

Christian Bokhove is Professor in Mathematics Education at Southampton Education School, University of Southampton, UK. He is also Director of Research at the School and has co-led their Mathematics, Science and Health Education research centre. He has a PhD from Utrecht University, Netherlands.

David Reynolds, CBE, was Deputy Pro Vice Chancellor, Head of Department and Professor of Education at Swansea University, UK. Since 2021, he has been Distinguished Professor of Education at Hangzhou Normal University, China. He has published extensively in the areas of educational effectiveness, comparative education and educational policy (over 20 books, 150 articles) and has lectured, consulted and/or researched in over 70 countries.

1

THE MASTER MATHEMATICS TEACHERS PROJECT

Chapter overview

- The beauty of mathematics and its impact
- Mathematics in primary schools: multiple aims
- Mathematics learning: multiple outcomes
- Mathematics teaching: multiple 'prescriptions'
- Mathematics teacher professional development
- Master mathematics teachers in China
- The MasterMT project: purposes
- The MasterMT project: the book

> Talent is important, but how one develops and nurtures it is even more so.
>
> – Terence Tao

We share the genius mathematician's view and believe that education is the crucial catalyser of intellectual accomplishment. In this book, we embark on a journey towards the masterful way of bringing out the mathematical talent from young children.

In this first chapter, we seek to give a grand view about: the beauty and impact of mathematics (1.1); mathematics as a subject in primary schools with the multiple aims (1.2); the multiple learning outcomes (1.3) and multiple teaching 'prescriptions' (1.4); mathematics teacher development (1.5); master mathematics teachers (MasterMT) in China (1.6); the MasterMT project (1.7) and the map of the book (1.8).

DOI: 10.4324/9781003127925-1

1.1 The beauty of mathematics and its impact

Mathematics, as the oldest academic discipline alongside philosophy (Krantz, 2010) and a discipline 'older than the oldest civilisations' (Boyer, 1968, p. 7), has been developed in various cultures separately for millennia and gradually integrated across the world over the past few hundred years (Boyer, 1968; Hodgkin, 2005).

In Moravia, evidence of number counting was found on a wolf bone dating back to 30,000 years ago (Cooke, 2013). In the ancient era of world's oldest civilisations, such as ancient Egypt, Babylonia (a.k.a. Mesopotamian), Greece, China and India, mathematics had already taken steady forms and been gradually accumulated and applied in human activities of various purposes (Merzbach & Boyer, 2011; Nuffield Foundation, 1994).

The renowned mathematician, Carl Friedrich Gauss, regards mathematics as 'the queen of all sciences'. Today, mathematics has never been more important in history than it is in the fast-changing times of ours. Not only is mathematics beautiful, it is also essential for many scientific developments in our society. Safe banking transactions would not be here if it weren't for cryptography and insight in prime numbers, Engineering and the aerospace industry builds on solid physics expressed in mathematical formulas, and we are able to peer far into the universe and analyse waves from far, far away with the help of mathematical models.

1.2 Mathematics in primary schools: multiple aims

Worldwide, mathematics is seen as an important school subject. The role of mathematics in everyday life and future work of learners has always been at the forefront of education reforms since last century, with terms and frameworks for mathematical competence gradually (re)invented and expanded over time (Geiger et al., 2015). Either numeracy or mathematical literacy expects the proficiency in mathematics beyond its abstract disciplinary sense; it expects competent learners to be able to model and tackle problems in seemingly non-mathematical contexts in profoundly mathematical ways. Despite a prevalence of research in Western, Educated, Industrial, Rich, and Democratic (WEIRD) samples (Muthukrishna et al., 2020), the role of primary schooling in many of the countries is similar. The foundations of later mathematical proficiency are laid in primary schools, with the procedural and conceptual knowledge taught there underpinning many skills we use to understand the mathematical world around us.

We see mathematics education as a similar vein: a broader discipline than just basic skills and procedural knowledge. However, it also is important to not see both as disconnected. Our study firmly starts from the starting point that the evidence on mathematics learning indicates that procedural knowledge supports conceptual knowledge, and vice versa. The relations between the two are bidirectional and reinforce each other (Rittle-Johnson et al., 2015). For the whole of mathematics education, so also for primary school level, we need to pay attention to both procedural and conceptual development of students. In the context of Chinese mathematics education, we can see a similar emphasis in the extension of what used to be the *Two Basics*, basic knowledge and basic skills, to the *Four Basics* by adding 'basic mathematical thinking' and 'basic mathematical activities' in the new curriculum standards initiated at the turn of the new millennium (MoE China, 2011).

In fact, scrolling through the curriculum standards for mathematics in countries with a wide range of cultures and traditions, you would find at least one thing in common: multiple aims. Take the mathematics curricula of America, China, England and Singapore as examples. To move away from a curriculum that is 'a mile wide and an inch deep', the Common Core State Standards for Mathematics (NGA & CCSSO, 2010) set eight goals for mathematical practice, from sense making in problem solving to reasoning independently as well as with others, from mathematical modelling to pattern identification and application of tools. In China, the Mathematics Curriculum Standards and its regular revisions (MoE China, 2001, 2011, 2022) expect learners to form four basics, four problem-based abilities (identifying, posing, analysing and solving problems) and ten core competencies (from number sense to mathematical reasoning and modelling, from the sense of symbols to the sense of application and creation). In particular, the latest version (MoE China, 2022) further refined the diverse aims, giving more consideration to core competencies, the coherence and consistency of learning between consecutive schooling phases, student self-regulation, creativity and affective outcomes. The National Curriculum in England (DfE UK, 2013) sets three broad aims for students in KS1–2: (1) becoming fluent in basic mathematics, (2) reasoning mathematically with high quality and (3) solving complex problems using mathematics. The Mathematical Syllabus of Singapore (MoE Singapore, 2012) expects primary school teachers to nurture learning in five dimensions: concepts (from abstract o concrete), skills (learning through understanding), processes (e.g., reasoning, communications, connection, and applications), metacognition and attitudes.

1.3 Mathematics learning: multiple outcomes

Building on existing theories and curricula, we expand the mathematics assessment framework that is conventionally almost exclusively spread around cognitive outcomes. In terms of mathematics learning outcomes, the MasterMT project seeks to measure three dimensions of mathematics learning – affective, metacognitive and cognitive outcomes. Given that these goals are already covered, albeit not in exact same terminologies, in various curricula of mathematics, including the Chinese mathematics curriculum, measuring them would yield evidence on implementation of such goals in master mathematics teachers' classrooms.

The concept of affective outcomes in mathematics is often termed as attitudes towards mathematics (ATM) indicating whether or not a student is happy about mathematics. Some found that ATM affects mathematics achievement (Ma & Kishor, 1997; Minato & Kamada, 1996), whereas others found mathematics achievement affects ATM (e.g., Garon-Carrier et al., 2016), which however does not change the fact that ATM is a partial result of mathematics education one receives. We have a more thorough discussion about previous research and quantitative findings regarding ATM from the current project in Chapter 5.

Metacognition, as another category of intrinsic competence, is defined as knowledge and regulation of cognition as one attempts to solve problems (Flavell, 1976, 1979). In education, increasing attention is given to the nurture and the measure of metacognition (e.g., Muijs & Bokhove, 2020; OECD, 2018), given its positive impact on cognitive outcomes in school subjects, such as mathematics (e.g., Desoete, Baten et al., 2019; Desoete, Roeyers et al., 2001; Kuzle, 2018; Ohtani & Hisasaka, 2018; Schneider & Artelt, 2010). In the field of mathematics education, mathematical thinking and problem solving is assessed often with cognitive-metacognitive models that consist of five factors (Lester et al., 1989; Schoenfeld, 1992): the knowledge base, problem-solving strategies, monitoring and control, beliefs and affects and instructional practices (as cited in Dossey, 2017, p. 66). Students with strong metacognitive skills are believed to be more easily to meet challenges in problem solving (Dossey, 2017). Cognitive performance in mathematics is found strongly predicted by metacognitive competence (e.g., Desoete, Baten et al., 2019; Desoete, Roeyers et al., 2001; Kuzle, 2018; Ohtani & Hisasaka, 2018; Schneider & Artelt, 2010). Given the importance of metacognition, our study looks at this in more detail, especially in Chapter 6.

In school mathematics, rational numbers and proportional reasoning are the most challenging topics in and of themselves (Lamon, 2007; NMAP, 2008). Students' competence of rational numbers and proportional reasoning predicts their performance in algebra and more advanced mathematics in secondary and higher education (Hurst & Cordes, 2018; Siegler et al., 2012). To estimate the cognitive learning outcomes that master teachers can nurture amongst their students, the MasterMT project gives a test on this most challenging part of primary mathematics, rational numbers and proportional reasoning, utilising the items that were regarded as hardest in the classical study in England, the Concept in Secondary Mathematics and Science (CSMS) project (Hart, Brown, Kerslake et al., 1985a, 1985b; Hart, Brown, Küchemann et al., 1981). Throughout the book, we use 'cognitive outcomes in mathematics' and 'mathematics performance/achievement' interchangeably. They refer to the focal outcomes that are measured with and inferred by the mathematics test given to the upper graders (i.e., those in Grades 5 and 6) amongst the project participants. Chapter 7 gives explicit discussion about existing research on rational numbers and proportional reasoning as well as quantitative findings in this regard from the current project.

1.4 Mathematics teaching: multiple 'prescriptions'

Despite potential influence of student background factors, researchers in education are always interested in one question above all: does education make a difference? Decades of educational effectiveness research indicates that teaching makes a bigger difference in learning outcomes than schools and systems (Muijs et al., 2014; Scheerens et al., 1989). It is even more so for the phase of primary education and in the case of mathematics (Luyten, 2003). A thorough study of classroom teaching is a crucial step towards understanding and improving teaching (Stigler & Hiebert, 1999). Previous studies have accumulated over time a number of teaching elements that contribute to better performance than expected.

The role of the teacher should not be underestimated. A similar conclusion follows from the Education Endowment Foundation's (EEF) evidence review on improving mathematics in Key Stages two and three (Hodgen et al., 2018). Many impactful approaches rely on teachers, for example their whole-classroom interaction and feedback (Miao & Reynolds, 2018). It therefore is hard to unpick the role of 'the teacher' without considering a broader set of effective strategies in the primary mathematics classroom, including the use of manipulatives and the sequencing of curriculum content. The teacher matters. Or rather, we should say *teaching* matters, which is supported by decades of teaching effectiveness research (Muijs et al., 2014), echoing one of the findings from the recent EEF evidence review (Hodgen et al., 2018, p. 15) that considers teaching quality 'important to the efficacy of almost every strategy that we have examined'.

Effective maths teachers spent more time *interacting* with the whole class so that all children are within the interactive teaching and learning 'radar' (Croll, 1996; Good & Grouws, 1979; Leung, 1992; Miao & Reynolds, 2018; Mortimore et al., 1988). In so doing, they are able to keep more children on task longer and more engaged in learning (Croll, 1996; Mortimore et al., 1988; Muijs & Reynolds, 2003). Effective maths teachers are able to manage the class well with little or no observable managing behaviours (Good & Grouws, 1977; Miao et al., 2015).

Highly effective maths teachers tend to emphasise the connectedness of mathematics (Askew et al., 1997). A better understanding of mathematical procedures and underpinning concepts can be promoted through building connections between multiple representations (Nistal et al., 2009; Noss et al., 1997; Schnotz & Bannert, 2003), between mathematical topics and ideas as well as between mathematics and the real world (Downton & Sullivan, 2017; García-García & Dolores-Flores, 2018; Sidney & Alibali, 2015).

In the Chinese maths class, it is students instead of the teacher who are at the centre (Mok, 2006). In fact, Chinese maths teachers tend to be free from the either-or dilemma between teacher-centred and child-centred (Huang & Leung, 2005). They would rather be the guide/guard who keeps children at the centre of learning and teaching by guiding/scaffolding their thinking through intensive chains of questioning-response-feedback to model students' thinking – typically metacognition (Miao & Reynolds, 2018). Such intensive questioning allows the formulation of *metacognitive experience* and the accumulation of *metacognitive knowledge* (Flavell, 1979) to reach all learners in the class. Analysing the Hong Kong portion of TIMSS Video Study data, Leung (2005) found more advanced content, more fully developed presentation, and more coherent, engaging and highly rated teaching in maths classes in Hong Kong where the class size is much larger than the case elsewhere.

Underpinning the quality of teaching as a key factor, in mathematics, is teacher knowledge. Teacher knowledge, and especially pedagogical content knowledge (PCK), is crucial in optimising the effective use of mathematics curriculum resources and interventions. Mathematics teachers differ from one another considerably in pedagogical knowledge and reasoning (Ball, 1988; Ma, 1999). About mathematics knowledge for teaching, Copur-Gencturk and Thacker (2021) found a disparity between teacher self-evaluation and actual assessment results. It suggests that the latter is more reliable than the former in accurately measuring mathematical knowledge for teaching. It is even more reliable to measure such knowledge in action, since it is through application of the knowledge in actual teaching that teacher knowledge truly functions in facilitating learning after all. Drawing on data from over 200 teachers and their students, Charalambous et al. (2020) found that whilst teacher knowledge contributes positively to student performance in mathematics, content knowledge and PCK converged into one dimension, instead of two. Teacher knowledge might be more of a connected whole rather than disentangled parts.

Whilst whether teacher knowledge is multidimensional or unidimensional remains a debate, it is widely accepted that effective mathematics teaching bridges learners and mathematics (Cohen et al., 2003; Hill & Chin, 2018; Shulman, 1986). Teaching is similar to clinical practice, and good teaching practice prescribes good learning experiences and results that suit the particular learners (Grossman & McDonald, 2008). In China, teachers' knowledge of curriculum standards, textbooks and students have been written in the new curricular standards since the 2001 curriculum reform (MoE China, 2001, 2011, 2022). It has subsequently become basic components in *Teaching Explanation* (说课), a nationwide standard practice carried out by teachers often in front of a panel of fellow teachers and experts before or after their lessons being observed. The conceptual structure underpinned the Mathematical Quality of Instruction (MQI) observation protocol (Hill & Ball, 2004; Learning Mathematics for Teaching, 2011) has long been widely regarded as valid ingredients for quality mathematics teaching in the practitioners' circle in China. In fact, these have been parts of standard practice.

1.5 Mathematics teacher professional development

To develop students' 21st-century skills – creativity, collaboration, communication and critical thinking, Beswick and Fraser (2019) argue that teachers need not only the essential knowledge but also the competence to collaborate with colleagues, communicate with peers and learn from each other. To improve the quality of teaching and teacher knowledge and competences, PD is key. Although evidence concerning specific effects of PD is limited, accumulated research evidence tends to suggest that extended PD is more likely to be effective than short courses.

Across nine studies, Yoon et al. (2007) found that teacher PD posed a moderate effect on students' achievement, accounting for about 21 percentile points difference between experiment and control groups. The review by Doğan and Adams (2018) confirmed the positive effects of teachers participating professional learning communities (PLCs) on student learning. Collaboration within the PLC leads to improved teaching effectiveness when such collaboration features instruction-focused activities (Cravens & Hunter, 2021). The policy review by Jensen et al. (2016) indicates that master teachers in Shanghai play a leading role in teachers' PD, for which they are funded, evaluated and promoted.

Teaching research groups (TRGs) are common forms of PLCs in Chinese schools (Cravens & Wang, 2017; Fan et al., 2015; Paine & Ma, 1993). Teachers within a TRG are led by their head of group who in turn works with the headteachers, particularly the headteacher in charge of teaching affairs. The main responsibility of a TRG is to:

- study curricular standards, textbooks and other teaching materials together;
- plan teaching and content at macro and micro levels, from the plan of the entire primary phase to that of a specific school year, a specific unit in a semester and a particular lesson;
- grade homework together and exchanging ideas;
- write, analyse and reflect upon test papers together;
- organise peer observations and feedbacks;
- prepare for teaching competitions together;
- plan and organise student teacher placement in collaboration with local initial teacher education institutions;
- plan and offer new teacher induction and mentoring.

(Fan, Miao et al., 2015; Paine & Ma, 1993; Wong, 2010)

Amongst all the purposes of the TRG-based PD, the central aims are to update ideas of teaching and learning, design new situations and improve classroom practice through exemplary lesson development (Huang & Bao, 2006) or public lessons (Han & Paine, 2010) for teaching research purposes. Like the Japanese lesson study, the Chinese way of lesson study is concerned with practical issues. In China, the TRG-based lesson study is feature with expertise developed over deliberate practice on focal topics (Huang et al., 2017). Lesson studies in both countries pay attention to developing a particular lesson through collaborative lesson planning, classroom observation and post-lesson discussion to tackle a particular issue. However, the Chinese lesson study process has a larger emphasis on the expertise stemming from experts, the revision of lesson plans and the subsequent new actions (Han & Huang, 2019; Huang & Bao, 2006).

In addition to the long established TRGs within schools, the Master Teacher Studio (MTS) -based PLC is also widely practised and regularly evaluated nowadays, supporting cross-school teacher development (Zhang et al., 2021). Both

TRGs and MTSs weave together practitioner-led research, professional learning and teaching improvement. As we will show in this book, Chinese master teachers are involved in intensive improvement of teaching on a regular basis within and across schools, but such consistent effort is not as simple as the assumed 'repeated' practice (Sims & Fletcher-Wood, 2021).

1.6 Master mathematics teachers in China

In fields such as medicine, generating best practice to the wider community is typical to the advancement of the field. In education, expert teachers have been found to be more effective in implementing curriculum and cultivating desirable learning outcomes than non-expert teachers (Berliner, 2001; Bond et al., 2000; Leinhardt, 1989).

In China, the official rank system promotes teachers from Third Level to Second Level, First Level, Senior, and then Professoriate-Senior ascendingly. The rank 'Professoriate-Senior' was not added to the professional rank system until 2015 before which there had been four ranks, ascendingly Third Level, Second Level, First Level, and Senior, since 1986. The newly added rank is equivalent to the rank of university professors, representing the top of the profession to this day. Professoriate-Senior teachers are nominated each year and the number of places is highly limited. In addition to professional ranks, teachers are also given honorary tiles, such as Super Teachers (特级教师), Subject Leaders (学科带头人) and Backbone Teachers (骨干教师), for excellent teaching practice and strong contribution to colleagues' PD, in addition to other merits (Fan et al., 2015; Huang et al., 2017; Li et al., 2011; Miao et al., 2022).

Master mathematics teachers are the cluster of teachers who demonstrate best practice in mathematics teaching and are well regarded by their professional peers locally or nationally (Cravens & Wang, 2017; Fan, Zhu et al., 2015; Zhang et al., 2021). Beyond outstanding teaching, they play a leading role in their peers' PD within and across schools (Fan, Zhu et al., 2015; Zhang et al., 2021). In the roll-out of reforms and PD, master teachers play an important role in PD activities such as the collaborative study of teaching material, collaborative lesson planning, peer observation and post-lesson reviews (Huang & Bao, 2006). Master teachers' views on what comprises effective teaching are deemed influential and important, with their conceptions about effective teaching bridging traditional features and innovative notions in teaching (Huang & Li, 2009). Prior research indicates that master teachers generally see an effective lesson as one with the following features:

- comprehensive and feasible teaching objective (knowledge, skill, mathematics thinking and attitudes);
- scientific and reasonable lesson design such as connections between content and the development of content;

- students' participation, self-exploratory learning, independent thinking, collaboration and exchange;
- teacher's sound subject knowledge and apt teaching skill, and good personality; and
- providing proper classroom exercise and homework as well as high-order thinking opportunities.

(Huang & Li, 2009, p. 299)

Within provinces, educational departments at municipal or provincial levels offer specific grants for master teachers to apply. Upon approval, they can set up a MTS over a three-year term and carry out a proposed professional project (Li et al., 2011; Zhang et al., 2021). In addition to the traditional TRG-based PD within schools, the MTSs make it possible for master teachers to extend their contribution to colleagues' PD across schools.

Though limited attention from research has been given to master teachers' practices, beliefs and their leading role in teacher PD in China, previous studies are relatively small scale and qualitative in nature. One original and important feature of the study in this book is that we study master mathematics teachers at scale. Seventy master mathematics teachers and 3,178 students participated in the study, hence providing a broader and wider insight into master teachers' practices and development in comparison with former studies.

1.7 MasterMT project: purposes

Collecting data from master maths teachers in China and their classes, the project focuses on three facets of maths teaching and learning: (1) *action*: the characteristics of master maths teachers' teaching and the kind of learning processes that they facilitate and students carry out in the maths class; (2) *mind*: the knowledge and beliefs teachers and students hold regarding the specific content taught and/or assessed and regarding mathematics in general; and (3) *pattern*: (3a) the big picture that defines where the master maths teachers are in the teaching profession and their students are in the learning of mathematics, (3b) the individual and collective effects of various factors at multiple levels on student achievement in higher-order knowledge, skills and thinking, their levels of metacognition, and their perceptions of schooling, mathematics learning and their maths teachers' teaching, and (3c) the professional trajectories through which teachers develop into master teachers.

In the pursuit of best teaching practice in primary maths, the project seeks to address three major research questions:

1. How well do master maths teachers teach and their students learn mathematics?
2. What do master maths teachers and their students think and know about teaching and learning of mathematics?

3. What are (a) the broad picture of master maths teachers and their students, (b) the potential teaching-learning mechanisms that explain affective, metacognitive and cognitive outcomes in mathematics and (c) general routes via which teachers grow into master teachers?

Under the guidance of the research questions, we have conducted multiple lines of enquiry, utilising a variety of qualitative and quantitative methods. In addition to the research questions, we have developed corresponding hypotheses to be presented and tested in Chapters 5–7 that focus on three types of learning outcomes (affective, metacognitive and cognitive) and the mechanisms behind excellent performance in each kind of learning outcomes.

1.8 MasterMT project: the book

This book is based on the MasterMT project. It is a tale of 70 master teachers and 3,178 students of which 3,033 take a mathematics survey on two or three of the focused domains, namely affective, metacognitive and cognitive outcomes in mathematics.

Chapter 2 presents a rich picture about a master teacher's work. It offers a long glimpse into the master teacher's work over 40 school days, including everyday classroom teaching, the quality of learning resulted in her class, her unique methods in developing children into reflective and self-regulated learners, and two teaching research meetings organised by the mathematics department of her school.

Chapter 3 illustrates the grand plan for the MasterMT project, getting us prepared for collecting data in master mathematics teachers' classes. Situating the project in the Integrated Paradigm, the chapter explains the data collection and analysis methods, the connections between methods and research purposes, and how the proposed findings may interconnect to address the research questions. More specifically, teaching is not only measured with three observation systems but also interpreted qualitatively; learning is measured with the Rasch model and interpreted both quantitively and qualitatively; teaching and learning mechanisms are explored with multilevel models and (multilevel) structural equation models; the analyses of teacher professional status and development draw a wide range of ethnographic data, including longitudinal case study data, teacher interview data, teachers' writing about their PD trajectories and the development of their master teacher studios, and our participatory observation of teacher PD in action in several teaching research events at the school, municipal, provincial and national levels.

Chapter 4 presents the patterns (QUAN) and characteristics (QUAL) of mathematics teaching across 70 master mathematics teachers' classes. Chapter 5 focuses on the patterns and variation of affective outcomes in mathematics and scrutinises the mechanisms that make students enjoy mathematics and the learning of it. Chapter 6 looks at metacognitive outcomes in mathematics and explores the mechanisms that nurture better knowledge and regulation of cognition amongst learners.

Chapter 7 captures the patterns and variation of cognitive outcomes in mathematics and models the mechanisms that cultivate strong learners in mathematics.

Chapter 8 listens deeply to teachers' views about the lessons we observed and their beliefs about mathematics teaching and learning in general. Through the deep listening, we gain insight into teachers' knowledge about mathematics, about mathematics teaching and about children and how they learn.

Chapter 9 captures the historical as well as the state-of-the-art pictures of teacher PD. The historical picture comes from master teachers' accounts of their journey towards masterly teaching and their reflection on the development of Master Teacher Studios. The state-of-the-art picture results in teacher PD in action as we carry out participatory observation of teaching research meetings/conferences at the school, municipal, provincial and national levels.

In Chapter 10 we conclude the project findings and discuss indications of lessons from master mathematics teachers for a better mathematics education for all primary schoolers. Beyond effective teaching is masterly teaching; above the science of teaching is teaching as an art.

2

40 DAYS IN A MASTER MATHEMATICS TEACHER'S WORK

Chapter overview

- Fraction masters nurtured by the master teacher
- The lesson in the master mind
- A long glimpse into a master teacher's work
- A head full of teaching ideas and creativity
- The resilient teacher: teaching by all means
- Keeping track of student progress
- Developing self-regulation amongst students
- TRG meeting 1 of 2
- A lesson on *Dividing a Fraction by a Whole Number*
- Meeting on the spot after the lesson
- TRG meeting 2 of 2
- Embarking on the nationwide search for masterly teaching

> To see a world in a grain of sand,
> And a heaven in a wild flower,
> Hold infinity in the palm of your hand,
> And eternity in an hour.
>
> – William Blake

Our journey into masterly teaching and learning begins with the participatory observation of Ms Q's work over 40 school days stretching two semesters in 2018.

This is an in-depth case study of a master mathematics teacher – Ms Q – a provincially renowned master teacher with more than 20 years of teaching experience.

DOI: 10.4324/9781003127925-2

She is currently the Deputy Head of Teaching Affairs Office and Head of Mathematics Teaching Research Group at her school. With multiple administrative responsibilities and numerous awards, she puts classroom teaching at the centre of her attention and teaches two classes every day as a regular classroom teacher. She currently teaches mathematics to two parallel classes in Grade 5 (at the same age as American Grade 5 and British Year 6) and has been teaching them from day one in Grade 1.

Taking the ethnographic approach, this chapter features the longitudinal case study of Ms Q's work over 40 days. We will have an in-depth understanding of: (1) the teacher's everyday teaching and the educational excellence she nurtures through teaching; (2) her beliefs in and passion for mathematics teaching and learning; (3) her strong will to teach against technical odds; (4) the student progress system she constructs; (5) her approach to developing self-regulated learners; and (6) her role in leading teacher PD from the bottom up through school-based teaching research groups.

To begin with, let's get into Ms Q's class and have a close look at a conclusive lesson that sees the transition of the 11-year-olds in her class from fraction *learners* to fraction *masters*.

2.1 Fraction masters nurtured by the master teacher

The class has just finished being taught the unit on Definition and Properties of Fractions. Today's session (15 May 2018) is a conclusion of the entire unit, *Definition and Properties of Fractions* (PEP, 2013, 5B, Unit 4).

The purpose is apparent – reviewing the whole unit through thinking, discussing and making notes together, at both the class and student levels. The teacher structured the lesson with an enormous number of questions – with one related to the next. Questions are interconnected mathematically and logically.

Ms Q tried to switch on the projector to see if there was any luck it would work again, but it didn't. Without hesitation, she announced the class had begun. As in every classroom in this country, the class started with the teacher and the class bowing to each other whilst exchanging greetings. The students often bowed deeper to show their respect.

The first thing Ms Q did was ask the class: 'What are the sections in this unit we have just learnt?'

Class: The Definition and Meaning of Fractions . . . Proper Fractions and Improper Fractions . . . Basic Properties of Fractions . . . Reduction of Fractions . . . Common Denominators . . . and Mutual Conversion of Fractions and Decimals. [answering randomly]

Ms Q: Who would like to tell us detailed aspects of knowledge that were covered in each section? For example, what have we learnt in the section, the Definition and Meaning of Fractions? . . .

[pause, looking around patiently and seemingly allowing some time for students to think]

... You [pointing at Student 1 (S$_1$)], please tell us.

S$_1$:   Fractions and Division.

Ms Q: Fractions and Division.

Ms Q counted one finger – a gesture indicating children to say some more so that she can carry on counting the number of subsection titles in the section.

This posture appeared to be mutually understood – the teacher was expecting the next answer from the class and she would carry on counting her fingers until every piece of information about the topic was put together by the class.

Ms Q appeared to be waiting with a thoughtful expression, looking to the rest of the class and indicating S$_1$ to sit down. She pointed to Student 2 (S$_2$).

Nobody looked things up in their textbooks. They were calling out the topic titles from memory.

Ms Q: You, please.

S$_2$:   *The Emergence of Fractions in History.*

Ms Q: *The Emergence of Fractions in History.*

         [saying to the whole class with a thoughtful expression]

         OK, look back and think about how fractions emerged?

The class answered randomly.

Ms Q: [echoing the class's random answers]

         Right. When we are dividing or measuring things, the result may not always be a whole number. We have to use fractions. Okay, in the section, Definition and Meaning of Fractions, we have learnt the Emergence of Fractions and then the Definition and Meaning of Fractions. What are details of the definition?

Class: Divide the unit evenly into a certain number of sections . . . [randomly answering]

Students called out all titles of each subsection in section 4.1, immediately after which the class became quiet.

Ms Q: Okay, what we have just talked regarding this unit's content appears to be fragmented, incomplete, and lacking details. This unit contains the largest amount of knowledge amongst all units in the textbook (PEP, 2014a) and it is harder to grasp and easier to make mistakes on than other units. So, it is crucial for us to review it systematically. In order to give the knowledge a more systematic summary, let's make the revision outline for this unit together.

Before the whole-class began to think together and each (including the teacher) write down a revision outline, Ms Q reminded the class of what they should be careful about in the process, such as turning the notebook to the location where opposite pages are both blank and offer a wide enough space for the notes.

Then she structured the remaining part of the lesson through Q and A (question and answer) with the whole class thinking, discussing, drawing conclusions and writing notes together. Textbooks were referred to with specific page ranges, detailed content and typical problem examples discussed at the class level.

At each step, when the appropriate answer emerged from the class, the teacher would repeat what the students said whilst jotting down the corresponding notes on the board. For some long definitions, the teacher repeated what students just answered and then wrote a long line to simply hold the space for what was just said. It seemed to be a convention that though the teacher omitted the entire writing of certain statements on the board like this, the students would never do so, as captured by the camera. They would always carefully write every single word in the notebook in an almost perfect fashion: the duty of learning is rightfully on the learners' shoulder.

The teacher is just there to guide/structure the learning, but the guidance seems so systematic and well-prepared. Children are, however, the centre of the classroom, since it is clearly their role to give all the answers to the questions that arise in the class. This is what Chinese educators termed the 'teacher-guided and child-centred' approach (Miao & Reynolds, 2018, p. 107). Everything seems well anticipated beforehand, which is not as easy as it appears to be. Such quality anticipation of what can be achieved in the class requires a sound understanding of the maths content, learners and maths teaching.

It is a universal phenomenon across the country in maths classrooms that the teacher makes every effort to say as little as possible about facts, solutions and answers. They would say several words – sometimes just one word or a gesture without saying anything – and leave the remaining part or all of a statement for the students to provide. They would frequently ask questions and expect the students to answer them.

The whole class gave their answers randomly. It was more of a whole-class discussion throughout the lesson. In fact, in all the previous ones, too. Children are so used to such whole-class brainstorming around questions constantly raised by the teacher and sometimes raised by their peers. Everybody's ideas flow naturally and are visible to everybody else in the class. The discussion is fully within the ability of the entire class. Individuals' thoughts are heard, not personally one-on-one with the teacher, but openly within the entire class. Thinking together almost all the time is featured at the class level. Students' individual contact with the teacher emerged only briefly during the individual seatwork (usually around 2 minutes) as the teacher quickly circulated through the class. Each volunteer during the whole-class Q & A seemed to have a one-on-one interaction with the teacher, but the rest of the class were apparently parts of the conversation, since they were all listening and thinking around the question and answers that were given by the volunteers.

What Ms Q and the class were talking and thinking about were basically: (1) what the unit offered in the textbook, (2) what the details of each aspect of knowledge were and (3) why they were defined in a certain way or expected to be applied in a certain way. What they were doing was thinking whilst writing down the corresponding notes on specific knowledge points that they just talked about or were talking about.

By the end of the lesson, the teacher completed her notes on the board and students did theirs in their notebooks (Figure 2.1). Typical problems had been given for the students to solve, but the teacher erased them to leave the key revision notes on the board.

One might imagine a maths lesson or the lessons of any other core subjects in a primary classroom without multimedia these days would mean that students are totally switched off, since children in this digital era are so used to all sorts of screens/interfaces. This is apparently not the case in Ms Q's class. Students seemingly have the learning habit that is built not magically within days, but years, beginning with their first lesson with the teacher in Grade 1. Their focus is on deeper thinking and the understanding of fundamental mathematics beneath the surface of numbers, representations and symbols.

At the end of the lesson, everybody becomes an expert on the topic, which is evident in their structured notes and diverse ways of representing interconnected knowledge in this area through pictures that they drew over the weekend (Figure 2.2). They have indeed become Fraction Masters. People in Chinese education like to quote Confucius: 'Acquire new knowledge whilst thinking over the old, and you may become a teacher of others.' Students in Ms Q's class are now ready to teach others fractions like a teacher.

Looking at these pictures, one could easily imagine how each learner animated in their mind their understanding of the knowledge, its structure and the connectedness of all elements in it, and then managed to map their imaginations onto the paper with pens/pencils/crayons, incorporating a personal element. Knowledge is hence both shared and personalised.

As the whole class worked to make a mind-map (Figure 2.1) for the knowledge learnt in the specific unit, the teacher applied various ways of questioning to scaffold the whole class's collective reflection on and summarisation of what had been covered over the past few weeks. Children were constantly offered opportunities to think out loud about intra- and interconnections between various knowledge points, mathematical procedures, concepts and sub-concepts. The teacher builds a series of mathematics problems to provide experiences of learning from concrete understanding to abstract reasoning, from typical examples to general knowledge points.

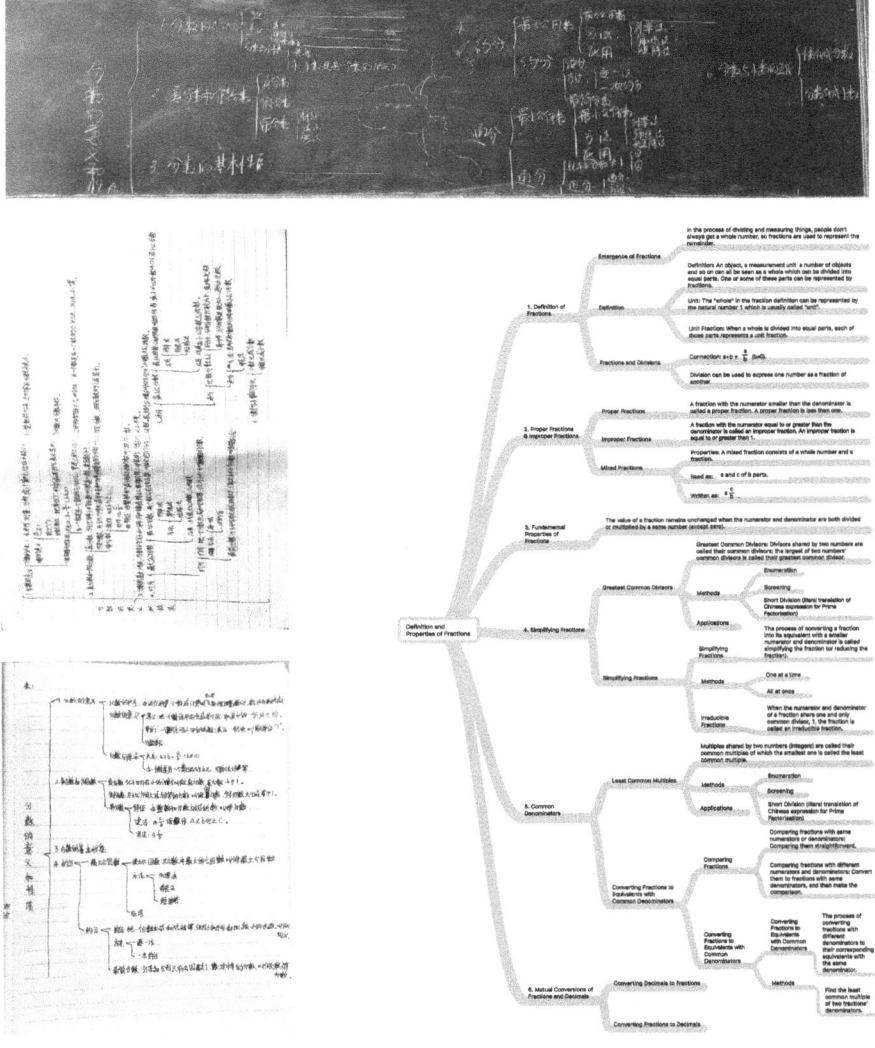

FIGURE 2.1　Ms Q's mind-map on the board and mind-map work samples from her class

2.2　The lesson in the master mind

The inclusion of metacognitive dialogue in the lesson echoes Ms Q's post-lesson reflection. In the interview, the teacher talks explicitly about (1) the purpose of each lesson step, (2) specific ways of relating knowledge in the unit to prior knowledge, (3) strategies for formulating certain cognitive and metacognitive experiences, and

a

b

c

d

e

f

g

h

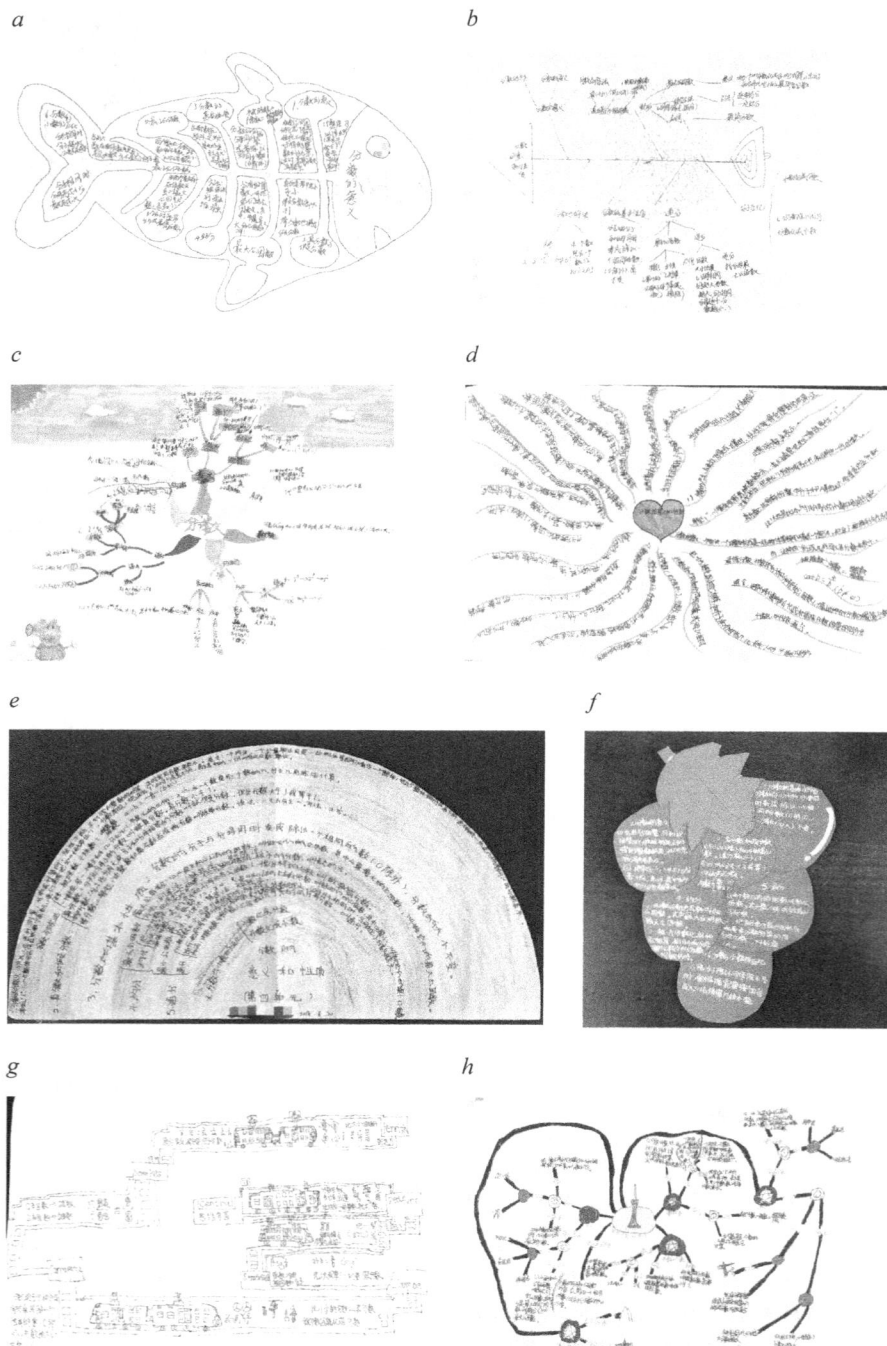

FIGURE 2.2   Personalised mind-maps redrawn over a weekend by Ms Q's students

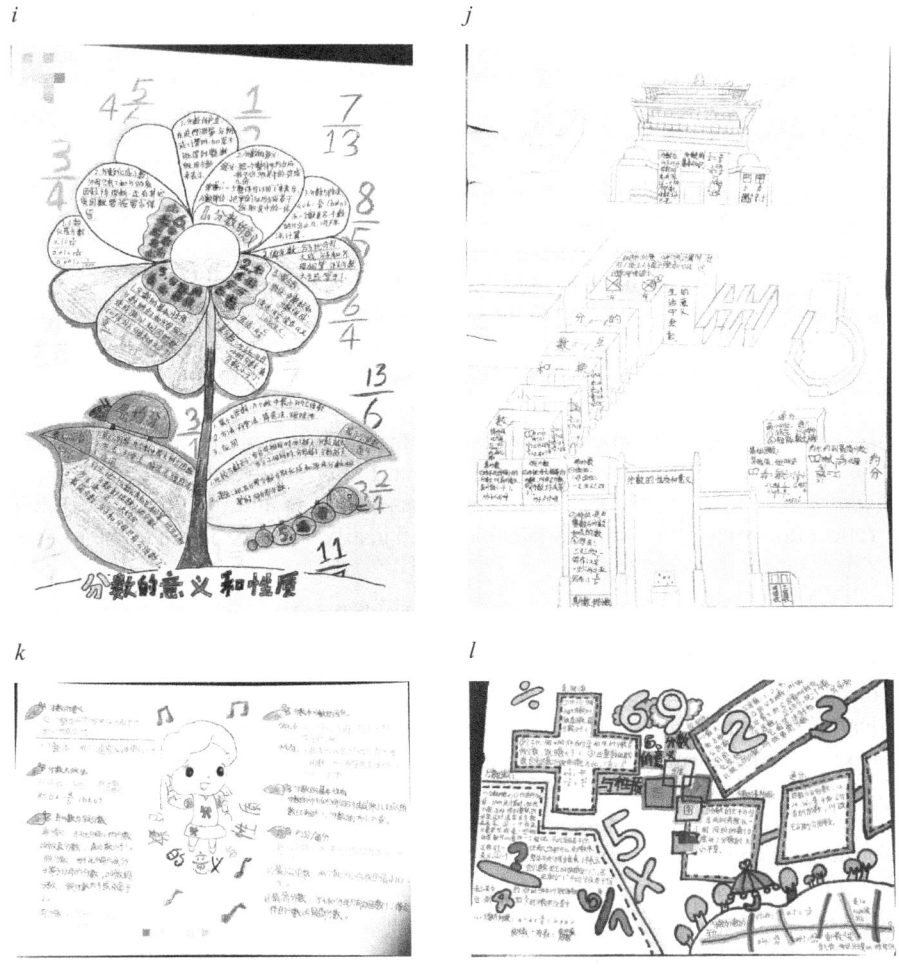

FIGURE 2.2   (Continued)

(4) interconnected points of knowledge within and beyond the specific unit on fractions.

Back at the office, Ms Q explains her idea on the lesson according to the mind-map that she constructed with the application called MindMaster on her laptop. According to Ms Q:

> The use of the spider mapping is to help students ultimately form knowledge structure in their mind. Every point of knowledge demands the children to form detailed statements.

(MasterMT, 20180515, itv401)

She clicks on the small plus sign to the right of the node titled *The Definition and Meaning of Fractions*. A statement instantly appears to the right: 'Divide one object or a group of objects into equal parts. One or a number of these parts are fractions.' She continues, clicking the node *Fractions and Division*. A statement appears: '$Dividend \div Divisor = \dfrac{Dividend}{Divisor}$, which can be represented with letters as $a \div b = \dfrac{a}{b}(b \neq 0)$.'

Because the IT facility was not functioning, it was not possible for Ms Q to play the electronic knowledge map to the class via the projector. The notes on the board tell the same thing, with the statements recalled and written down by the students, which seems to work better since the class have more opportunities to activate and retrieve learnt knowledge from their memory.

Across Ms Q's lessons, the most typical feature is that great emphasis is placed on cognitive and metacognitive development. In her view, students should have a complete mastery of the key concepts and skills in the unit; they must be able to reflect upon what they know and explain and justify the knowledge to their peers.

2.3   A long glimpse into a master teacher's work

Combining quantitative and qualitative evidence from this master maths teacher's classroom, the chapter sheds light on connections between classroom teaching, learners' metacognitive development and teacher metacognitive accounts of maths teaching and learning strategies. Detailed results and findings contribute to our understanding as to what makes profound mathematics teaching and learning in the primary school. Data in this section partially come from class observations (both structured and unstructured) and post-lesson teacher interviews.

Figure 2.3 shows the timetable for one of the two Grade 5 classes that Ms Q teaches. Every school day, Ms Q only teaches one maths lesson in each class, with subject-specialist teachers teaching other subjects, such as Chinese, English, Music, PE and Science.

In this class consisting of 48 students, we video-recorded 40 lessons that were mostly consecutive. The first consecutive lesson series (14 lessons) covered *Definition and Properties of Fractions*. We use this series of lessons to illustrate the structure of learning and teaching in the class. The topics are:

- L1 – *The Origin & Meanings of Fractions* (23 Apr 2018);
- L2 – *Fractions & Division I* (L2, 24 Apr 2018);
- L3 – *Basic Properties of Fractions* (25 Apr 2018);
- L4 – *Fractions & Division II: One Number as a Fraction of Another* (26 Apr 2018);
- L5 – *Proper Fractions & Improper Fractions* (27 Apr 2018);

Timetable of Class 5(1) in 2017–2018 (Translated by the first author from Chinese to English)						
		Monday	Tuesday	Wednesday	Thursday	Friday
AM	08:30–08:48	Morning Exercises				
	1	08:50–09:30	Chinese	Maths	Chinese	Maths	Chinese
	2	09:40–10:20	Maths	Chinese	Maths	Chinese	Maths
	10:20–10:28	Eye Exercises				
	3	10:30–11:10	English	Chinese	Reading & Writing (Chinese)	Arts	English
	4	11:20–12:00	PE	Mental Health	PE	IT	Music
PM		Lunch Break				
	5	02:00–02:40	Young Pioneer Activity	Arts	English	PE	Moral Education
	6	02:50–03:30	Science	Maths	Calligraphy	Music	Science
Please stick strictly to the timetable. Should any change of plan arise, please inform the Teaching Affairs Department immediately.						

FIGURE 2.3   Timetable for fifth graders in Ms Q's class

- L6 – *Highest Common Factors (HCFs)* (2 May 2018);
- L7 – *Using HCFs to Solve Real-World Problems* (3 May 2018);
- L8 – *Simplifying Fractions* (4 May 2018);
- L9 – *Simplifying Fractions & HCFs: Exercise* (7 May 2018);
- L10 – *Lowest Common Multiples (LCMs)* (8 May 2018);
- L11 – *Applying LCMs to Solve Real-World Problems* (9 May 2018);
- L12 – *Reduction of Fractions to the Same Denominators* (10 May 2018);
- L13 – *Mutual Conversion between Decimals and Fractions* (11 May 2018);
- L14 – *Definition & Properties of Fractions – Revision* (15 May 2018).

There existed a tendency of the teacher to spend a larger proportion of time on whole-class interaction in lessons, introducing new concepts for the first time than in lessons that followed the 'ice-breaking' lessons (NC1-L2 to L14 in Figure 2.4). This is observable in the first lesson, *The Origin and Meanings of Fractions*, where the teacher allocated 93.5% of the lesson time for whole-class interaction.

a

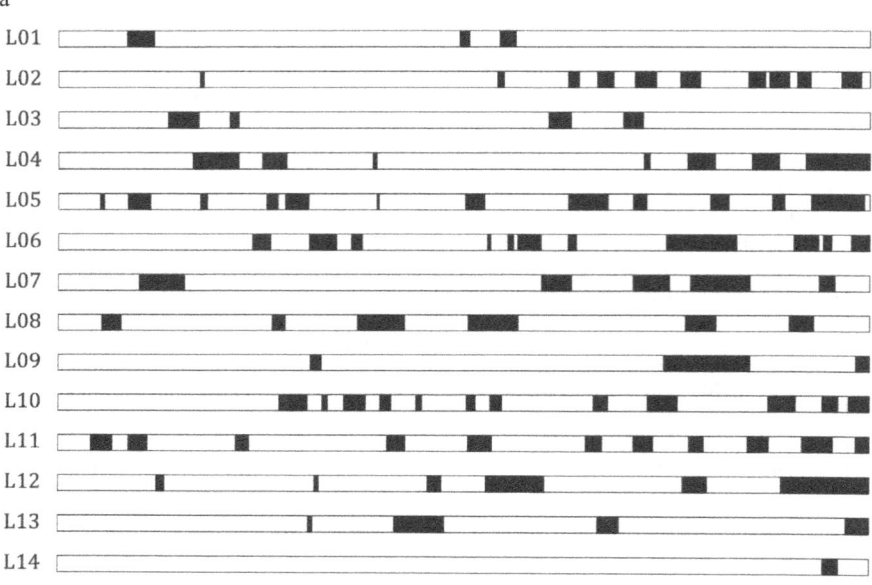

b

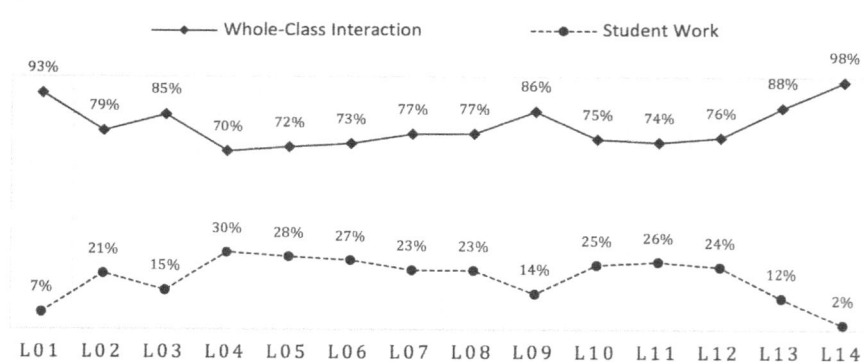

c

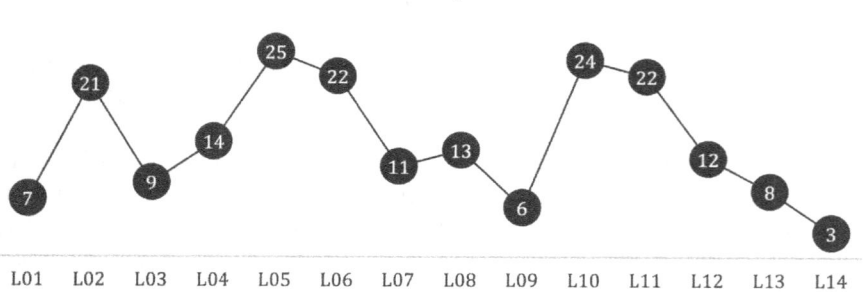

FIGURE 2.4 OTL segments in the 14 consecutive lessons by Ms Q

Note: In part a of the figure, White = whole-class interaction. Black = independent/group work. From left to right, the sections on each bar represents the OTL activities taking place in a lesson sequentially.

A large proportion of lesson time is spent on whole-class interactive teaching in which the teacher scaffolds children's mathematical thinking through intensive question-driven conversation. Such question-driven conversation is purely about the intended mathematics and its applications. It forms a consistent line that runs through every lesson. The question-driven conversation has multiple functions, including but not limited to the following aspects: (1) scaffolding learning, (2) facilitating active thinking, (3) promoting the development of cognitive and meta-cognitive skills and (4) assessing student cognition formatively.

There is a small proportion of lesson time on independent work or paired group-work/discussion (OTL-3, student work). Each segment of student work often lasts no more than one minute.

Overall, in lessons 1 to 14, there are a varying number of OTL segments (Figure 2.4). The form of the lesson shifts frequently, which makes the lesson pace brisk and children engaged. All lessons started in whole-class interaction and all but one also ended in whole-class interaction. It is surprising that all segments in each lesson occur in an exact pattern – whole-class interaction – independent/group work – . . . – independent/group work – whole-class interaction. Whole-class interactions and independent/group work occur in turn.

Another typical feature of Ms Q's teaching is the use of multiple representations on a same concept. In the sequence of lessons on fractions, the teacher shifts frequently between area model, number line model and concrete model of fractions which are at certain points explicitly related to division and decimals as well. This way, children are able to master the same concept based on thorough understandings.

In and across lessons, concepts are gradually developed. In all lessons, the teacher poses a series of mathematics problems which open up a rich pack of learning: from concrete understanding to abstract reasoning and then back to solving concrete issues (real-world/word problems); from typical examples to general knowledge points and then back to specific problems with applicable knowledge. Throughout all the lessons, the entire class are actively thinking about mathematics. Other activities, such as random social chats, are not observable. Almost all students are on task; on rare occasions where one or two students briefly drift off task, they are instantly spotted and reminded by the teacher. The average time on task across lessons is 99%. The steady effort from the students is striking. The learning habit is already there, so the class do not need exciting activities or games to get themselves engaged. It seems to be natural instinct for them to make continual effort and pay full attention to what is going on in the class. However, students' automatic effort in actively joining in the lesson should be attributed to their teacher's cultivation and reinforcement over a long term.

2.4 A head full of teaching ideas and creativity

There seemed to be continual teaching ideas popping up in the teacher's mind. Ms Q said every now and then an idea of a better way of teaching or presenting a specific knowledge point came to her, sometimes when she had only just woken up. She attributes the reason for her grey hair to the fact that she constantly thinks about designing lessons and PowerPoints slides to go with them. Her mind never seems to rest:

> My partner is constantly astonished to see my obsession with teaching. I often rush to work on my computer as soon as I wake up, with some new ideas for teaching in mind.

For Ms Q, ideas do not just keep popping up every now and then, they are carried out and experimented with in the class. The last semester (spring 2018) has come to an end right before the summer heat, and then the school commences its two-month summer break. Early September sees the beginning of a new semester and the change of seating plan in Ms Q's class. Previously the class was organised in six rows of eight seats: it is still six by eight, but now they sit in eight circles forming eight groups of six. Each group has given themselves a fancy name. As the class is known as the Oceanic Family, all group names are oceanic: Team Nimmo, Team Whales' Dreams, Team Turtles, Team Sharks' Speed, Team Starfish's Army, Team Fortune Crabs, Jumping the Dragon Gate Team, and Team from the Ocean Floor (Figure 2.5). Half a wall of the classroom is dedicated to the Competition of the Oceanic Family, displaying a list of the groups and showcasing the credits each group achieves.

Upon our return to fieldwork in Ms Q's class in September 2018, she excitedly explains the new seating plans that are being experimented with in her two classes. Although there are always mixed-ability groups of two, four and more for different class activities, she feels that seating in rows somehow prevents deeper collaboration between students. This time, she wants the mixed-ability grouping to really promote peer learning and boost an optimal level of competitiveness and collaboration between groups and students.

2.5   The resilient teacher: teaching by all means

In May 2018 when the IT system broke down halfway through a lesson, Ms Q kept teaching in a solid, effective way. After the projector and the computer stopped working in both classes, Ms Q continued to deliver her daily lessons to both classes. The absence of modern technology does not affect teaching pace or compromise the quality of teaching and learning in her class.

Besides, Ms Q has accumulated in the drawer of her desk all kinds of teaching and learning tools made by herself for different topics.

With shopping bags as packages, recycled sweet boxes are used as containers of the teaching and learning tools made by hand. The rest of the material involves paperboard as well as self-made pictures and shapes. Some of them look specific, for example the band of ten sticks drawn on the paperboard is apparently for the

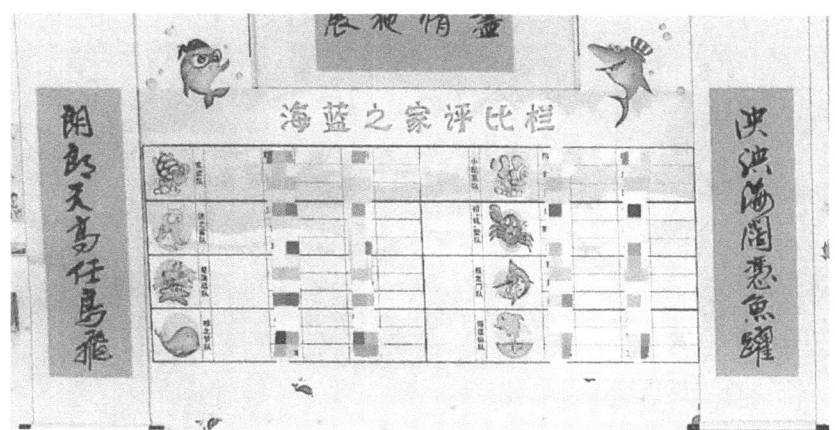

FIGURE 2.5   The ocean-themed grouping (mixed ability) in Ms Q's class

Note: Student names are blurred by authors for confidentiality.

teaching of place value. Some of them, such as the circles cut from paperboard (Figure 2.6: b), look normal, but can be used creatively for a lesson on a specific topic. On 24 April 2018, three of these circles appeared in the lesson on the *Relationship between Fractions and Division* (PEP, 2014a, p. 49), representing mooncakes to be sectioned evenly into four parts by the class and their teacher.

The first task in the lesson asks about the number of mooncakes that one person would have if four people were to share three mooncakes equally between them.

a   b   c

FIGURE 2.6 Hidden treasure in the draw of Ms Q's desk – hand-made teaching and learning tools

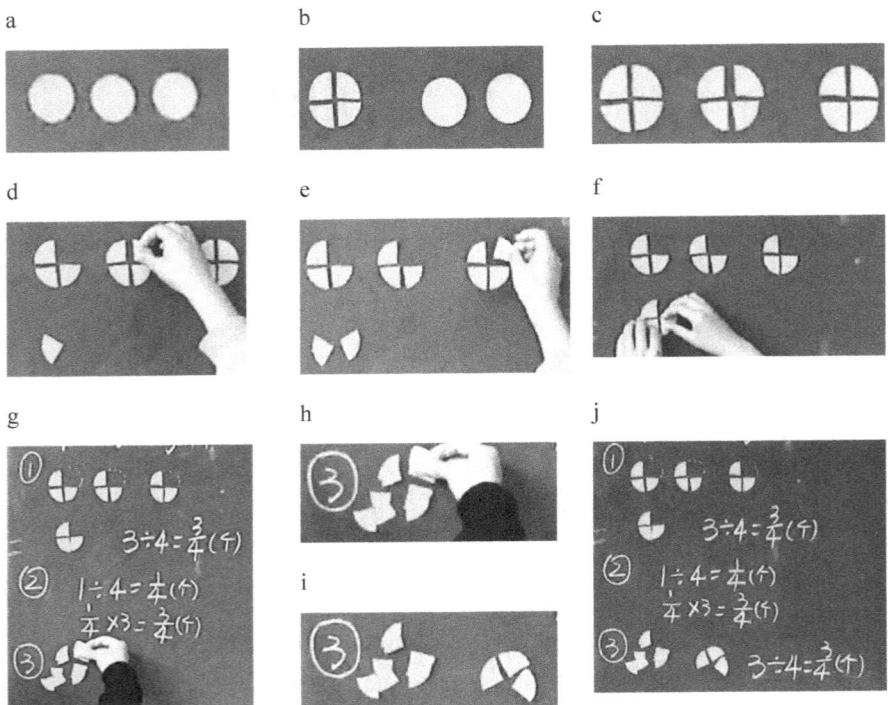

FIGURE 2.7 Using hand-made magnet 'mooncakes' to represent fractions in Ms Q's class

Three methods are raised by the students. With the second method being arithmetic, the first and third methods involve the even partitioning process in which Ms Q cuts the three paperboard circles accordingly to synchronise student articulation of the process (Figure 2.7).

The first method is explained by a boy who says that each mooncake can be evenly divided into four sections and every person gets one section of each mooncake – which is one-fourth – and ends up with three-fourths (Figure 2.7: a–c). Ms Q asks him to pick the sections from the partitioned 'mooncakes' on the board and put them down below to represent the overall sections a person gets (Figure 2.7: d–f). The third method (Figure 2.7: g–j), cutting the entire pile of mooncakes evenly into four sections, results in the class responding to Ms Q's question, 'Is there another way of cutting the cakes?' After the cutting, it becomes clear that the three mooncakes are now split into four piles, each consisting of three-fourths – the exact number of portions allotted per person. Ms Q puts it on the right-hand side to allow the class to see the three pieces of paperboard clearly and asks, 'What does this represent?' The class responds immediately, 'Three-quarters.' She writes down the number sentence to the right, asking, 'So, this means 3 divided by 4 equals . . . ?' 'Three-quarters,' says the class. The numerical representation of the third method now appears to the right of the 'mooncakes'.

The maths lesson on each school day is carefully planned for a specific content. Since both the total amount of time in a semester and that of maths content expected to be taught are fixed, teachers in this country are very cautious about sticking to the timetable. Chalk and board, projector and interactive whiteboard are just superficial tools if they are not utilised effectively to mediate the teaching and learning of the subject. What really matters is how teachers teach to develop children's understanding of fundamental mathematics. In this sense, ideally given optimal teaching, any tools could work. This echoes what the provincial teaching research official, Mr S, once emphasised to an audience of primary maths teachers in a provincial teaching research conference: in order to be truly proficient in teaching maths, teachers should perfect their fundamental teaching skills and be less reliant on the technology. Otherwise, he worries that some of the young teachers may find it hard to teach without PowerPoint.

2.6   Keeping track of student progress

Ms Q pulls out a big pile of A4-sized documents. These are the records that she has made over the past five years about the learning progress made by the Grade 5 children in her class (Figure 2.8). She has their homework performance as well as their test performance carefully documented. The records are not only made comprehensively in a printed table but also detailed alongside each specific maths problem where numbers are written: the blue numbers represent class codes of students who did not get the problem right in class A and red in class B.

She also shares with us details about the enriched homework that students were given over the past Chinese New Year holiday. In addition to the conventional booklet for winter holiday that is typical across the country, an alternative homework was given: (1) a 'mind-map' on all the mathematical errors made over the past semester; (2) a story or picture book entitled 'Magic Cuboids'; and (3) a piece of written work entitled 'My Maths Hero' based on an interview or 'I Am a Financial Expert' based on personal experience.

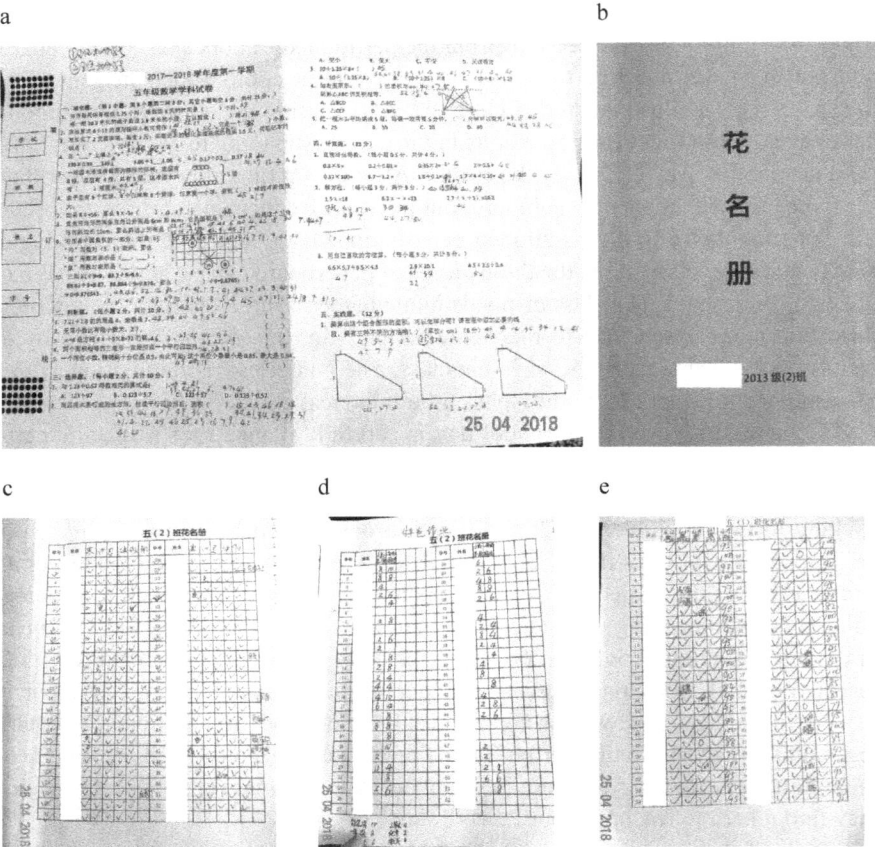

FIGURE 2.8 Tracking student progress in various dimensions – Ms Q's 'master plan'

2.7 Developing self-regulation amongst students

Ms Q asks students to write about their progress. This has apparently become a habit whereby children are able to monitor their own learning. The widely appealed competence, self-regulation, seems already a reality in the class.

A pile of notebooks sits on the corner of her desk. These are called the Collection of Errors. Each student has one. We pick up a dozen of them and quickly scan through them (Figure 2.9). To our astonishment, children's attitudes shine through the lines they write. You can see that they take these errors very seriously. They are not satisfied in simply correcting the errors. They copy each problem that they did wrong, show both correct and wrong answers/solutions, and analyse why they got it wrong. Some even make their own conclusions of their errors. Below this there are parents' comments and signatures, too. It is certainly not an intrinsic habit. It is

a

b

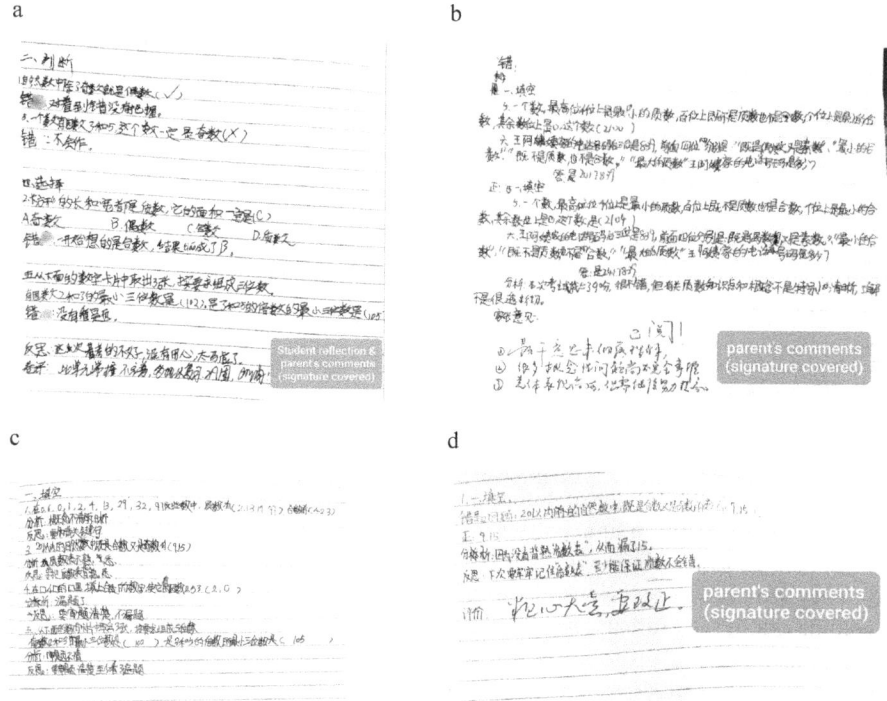

c

d

FIGURE 2.9   Self-regulation notes samples from Ms Q's class

a result of teacher expectation over time, but from the content one can tell children take it seriously.

Our participatory observation helps shed some light on a master teacher's work over the 40 school days and the natural influence she made by teaching in a masterly manner. Next, we complete the final part of our observational trip to Ms Q's school, taking part in two teaching research group (TRG) meetings where Ms Q taught a demonstration lesson (meeting 1) and gave a presentation on the use of ICT and, more specifically, the smartboard for classroom teaching (meeting 2).

2.8   TRG meeting 1 of 2

TRGs are professional learning communities universally existing in Chinese schools. Teachers of the same subjects, sometimes the same grades, meet regularly – often weekly – to discuss teaching related issues. The activities mostly involve observing and commenting on each other's lessons. This is the same case as in Ms Q's school. In the remainder of the chapter, we invite you to join us in observing two TRG meetings of the mathematics department at Ms Q's school.

As the deputy head of the teaching affairs office, Ms Q leads the TRG to study and plan lessons together along with the head of the teaching affairs office, Mr L. Their meeting is scheduled weekly or biweekly but can be flexible in terms of specific timing. The first meeting we observed is featured with a demonstration lesson delivered by Ms Q on 28 September 2018. At the end of this meeting, Mr L proposed the theme of the next meeting should be a tutorial on the use of the school's newly equipped smartboard by Ms Q since her demonstration lesson was eye-opening to everyone in terms of smartboard use. We thus followed them to observe this TRG meeting scheduled two weeks after, on 12 October 2018. Now, let us be prepared for the first meeting.

2.9 Prelude

A team of observers sitting in. The municipal teaching research official, a former maths teacher with decades of teaching experience, is invited to join the teaching research event in Ms Q's class. Before the lesson is delivered, a team of around 20 teachers, including the municipal teaching research officials, are seated in three rows at the rear of the classroom. The children look relaxed and ready for a normal day. They are used to occasions like this, having been to greater events, hence this is nothing out of the ordinary. As a matter of fact, Ms Q's class has been frequently visited by all sorts of visitors – teachers from other parts of the province, senior students from the university on teaching placement, colleagues and leaders from the school, teaching research officials from the city or province, researchers like us.

Teaching with Oceanic Themes. It is not just the grouping and class culture that we discussed earlier that is oceanic – the entire lesson to be observed today is set within an oceanic theme. The warming-up exercises (a whole number divided by another whole number vs the whole number multiplied by a fraction – the reciprocal of the other whole number) on prior knowledge are followed by a question raised by a cartoon fish, asking, 'What did you find?' The unpacking of the key task for the lesson is performed by another fish cartoon swimming from right to left: 'Divide 4/5 of a piece of paper into two equal parts. What fraction of the paper is one of these parts equivalent to?' The four tasks for assessing and challenging newly learnt knowledge and skills are all featured with oceanic creatures or expressions.

2.10 A lesson on *Dividing a Fraction by a Whole Number*

This lesson, on *Dividing a Fraction by a Whole Number*, lasts 41 minutes and consists of four major parts. The first part (2 min 36 sec) sees the activation of students' prior knowledge. The second part (24 min 32 sec) develops new knowledge amongst learners by offering them opportunities to explore, discuss and reflect. The third part sees the consolidation of newly learnt knowledge and the teacher's formative assessment of learners' knowledge status with a rich collection of problems (12 min 1 sec). The last part reminds the class to look back and summarise the new knowledge both verbally and algebraically, which takes just 1 min 53 sec since there have been thorough summaries throughout the lesson.

2.10.1 Part 1 of 4: Activating prior knowledge (00:00–02:36)

After the teacher and the class greet each other (00:08–00:21), the teacher starts the lesson by asking the class to recall what they learned yesterday. The class say it out loud, almost simultaneously: 'The Reciprocal!'

Then the teacher turns to the next slide which appears from the right-hand side with a problem that reads, 'Please point out the reciprocals of the following numbers: 3/10, 8, 7/2, 1, 1/4, 0.'

A group of volunteer students are invited to say out loud, quickly, the reciprocals. The reason why 0 does not have a reciprocal is discussed at class level: there is no such number for 0 to multiply with and result in a product of 1.

This is quickly followed by the second series of tasks shown on the slide: six tasks in three rows and two columns. The left column is about dividing a whole number by another whole number; the right column is about multiplying the whole number by the reciprocal of the other whole number. The children are able to calculate both tasks and see that the answers to both are equal. The equality between the left and right tasks is in fact the starting point of today's lesson where the dividend is going to be a fraction instead of a whole number.

2.10.2 Part 2 of 4: Developing new knowledge through exploration, discussion and reflection (02:36–27:08)

(2a) Thorough exploration into the first and key task

Upon completing the interaction around the two series of tasks, Ms Q says, 'OK, this is what we learnt previously. So, what we are going to learn today?' At this point, she knocks on the board to show the next slide. A fish cartoon swims quickly from right to left, showing the problem:

> If 4/5 of a piece of paper is divided into two equal parts, what fraction of the paper does each part represent? (Translated Task 3, Figure 2.10d)

Ms Q says, 'Look! The fish is coming. Read the problem quickly. Who can tell me the mathematical information that you have got?'

Through questioning, Ms Q gets the class to talk about the key information embedded in this problem and the unknown, whilst marking the information that the students point out with a red pen.

She then asks the class to recall in which year they learned equal sharing. Students immediately recalled that it was Grade 2.

Ms Q:	Here, what is equally divided into two parts?
Class:	4/5 of a piece of paper.
Ms Q:	What is equally divided into two parts? [repeating to prompt thinking]
Class	4/5 of a piece of paper. [more affirmatively]
Ms Q:	So how do you represent 4/5 of a piece of paper?

Some raise their hands high in the air. *Ms Q asks S$_1$ to answer.*

S$_1$:	Divide a piece of paper equally into five parts and pick four of the parts.
Ms Q:	Okay, you mean dividing a piece of paper into five equal parts.
	[responding thoughtfully whilst tapping on the smartboard where an area model of 5/5 is shown]
	[turning the discourse to the class]
	Right?
Class:	Yes!
Ms Q:	Pick from them . . . ? [in prompting tone]
Class:	Four parts.
Ms Q:	[tapping on the board to show an animation of the four parts being quickly shaded]
	Then what?

With Ms Q seemingly pausing her questioning, some students put up hands again, indicating they have answers.

Ms Q:	Having picked four parts, what is the next thing to do?

More students are raising their hands.

Ms Q:	[pointing at S$_2$]
	Come on. You.
S$_2$:	Divide into two equal parts.
Ms Q:	Divide into two equal parts. Divide what into two equal parts?
Ms Q:	Those of you raising hands, say it together. Divide what into two equal parts?
Class:	Divide 4/5 of a piece of paper into two equal parts.
Ms Q:	How do we express it mathematically?
Class:	Divide 4/5 by 2. [saying it out loud]

Perhaps for extra emphasis, Ms Q quickly points at S$_3$ to the right of the class, inviting him to say it.

S$_3$:	Divide 4/5 by 2.

Ms Q: [tapping on the smartboard to show the arithmetic expression]
           In what way is this expression different from expressions we learnt previously?

With many raising their hands, Ms Q invites S_4 in the middle of the class to answer it.

S_4:   What we have learnt is dividing a whole number by a whole number.
Ms Q: Dividing a whole number by a whole number. Right.
           [nodding head]
S_4:   But now what we have here is fraction as the divisor.
Ms Q: [immediately saying in an upwards tone with a big smile]
           Fraction as the *divisor*?
Many & S_4: Fraction as the *dividend*!
Ms Q: Fraction as the dividend!
           [pointing to problem on the board briefly and picking up a magnet card]
           So, that is to say, what we are learning is . . . ?
           [in an upwards tone, looking at the class]
Class: Dividing a fraction by a whole number.

Ms Q sticks on the top left of the chalkboard the magnet card where the lesson title (Dividing a Fraction by a Whole Number) is printed.

Ms Q: Maybe some of you already know the answer. Who could tell me the answer out loud?

Many raise their hands high in the air.

Ms Q: Those raising hands, say it together!
Class: 2/5.
Ms Q: 2/5. Okay. The 2/5 that you said. Do you know why it is 2/5?

Many are still raising their hands.

Ms Q: Okay, there is no need to rush. As the saying goes, truth comes from practice.
           [turning to the smartboard and tapping it to show the seatwork explanation]
           So, you will need to represent what?
Class: 4/5.
Ms Q: And then you will need to represent dividing it (that is, 4/5) . . . ?
Class: . . . equally into two parts.
           [completing the teacher's statement]

Then, re-emphasising the seatwork explanation on the screen briefly, Ms Q gets the clock set for 2 minutes, 'OK, ready, go!'
Everyone starts to work on the task. Ms Q circulates, observing everyone's progress carefully without drawing attention.

She walks through the class as students are busy folding and shading paper to represent the way to divide 4/5 by 2.

(2b)   The emergence of two methods and multiple representations

The timer soon rings and the class show their methods produced during the seat-work time which Ms Q had already accurately anticipated. Whilst 'cruising' around, she has collected two samples of students' work. Quickly taking a picture of each with the camera of the smartboard, Ms Q lines them up on the screen.

Representing two methods for task 1 with area models

The 'author' of the first method is invited to talk the class through the rationale behind his representation of $4/5 \div 2$. The student comes forward and shows the class his method.

He first divides the paper into five equal parts and shades four parts of it. Then he folds the 4/5 horizontally in half to get the answer.

Then the teacher asks the class to guess how the author of the second method shown on the smartboard did it. The answer comes quickly: further folding the five-fifths of the paper vertically results in dividing it into ten equal parts. Now, a half of the $\dfrac{4}{5}$ or $\dfrac{8}{10}$ is clearly $\dfrac{4}{10}$.

Arithmetic representations of two methods for task 1

Having understood the process by redoing the folding, the teacher suggests the class summarise the two methods. She clicks on the board to get the animation shown bit by bit as she continues interacting with the class by posing questions.

The teacher guides the whole-class discourse towards representing the processes arithmetically as such:

$$\frac{4}{5} \div 2 = \frac{4 \div 2}{5} = \frac{2}{5}$$

$$\frac{4}{5} \div 2 = \frac{4}{5} \times \frac{1}{2} = \frac{4}{5 \times 2} = \frac{2}{5}$$

The rationale of the last method is explicitly reasoned and built on closely inter-connected representations, verbal explanations, arithmetic processes as well as graphics – the area model (Figure 2.10d).

(2c) Seeking the more generalisable method out of the two

Two methods are compared with the whole class quickly calling out the answers to a series (#4) of three tasks with varying divisors, shown on the slide one at a time:

$\dfrac{6}{7} \div 6$, $\dfrac{6}{7} \div 3$, $\dfrac{6}{7} \div 2$. Then, the next slide shows a task (#5) that is different from these three: $\frac{4}{5} \div 3$. Everybody hesitates for a second and suddenly laughs at themselves because they have lost the rhythm built on the former tasks.

The sudden contrast of difficulty between two slides leads the teacher to suggest the class compare the two methods on the board. Through answering the teacher's questions, the class make it clear that the first method, suiting the former three tasks, has its limitations, whereas the second method can be more widely applied. At this point, the class is asked to find out the challenging task on page 30 of the textbook and work it out. After the seatwork period of time (16:53–17:36), the teacher initiates a class-level discussion on the calculation process of the task.

Pointing at 4/5 ÷ 3 on the smartboard and then 4/5 ÷ 2 on the left of the chalkboard (Figure 2.10e), Ms Q asks: 'Looking at this problem and that problem, what did you find about the law of dividing a fraction by a whole number?'

This task offers students more opportunities to try out the more generalisable method – dividing a fraction by a whole number is the same as multiplying the fraction by the reciprocal of the whole number arithmetically, with the area model to enhance understanding of the arithmetical processes. This step further prepares students for a more formal statement of the arithmetic solution in the next step.

(2d)   Summarising the law of dividing a fraction by a whole number with timely consolidation

As the teacher poses a series of questions, verbal, symbolic and graphical representations for solving '4/5 ÷ 3' emerge from the whole-class discussion. The class thus come up with a more generalisable solution for dividing a fraction by a whole number that is nonzero: multiplying the fraction by the reciprocal of this whole number. This generalised law of dividing a fraction by a whole number is expressed verbally by the students and then added to the chalkboard by the teacher. This method always works, whether or not the numerator of the dividend is a multiple of the divisor.

With the calculation law stated and defined like a theorem, the teacher asks the students to try it out with two more tasks on page 30:

$$\frac{9}{10} \div 3 = \frac{\square}{\square} \times \frac{\square}{\square} = \frac{\square}{\square} \qquad\qquad \frac{3}{8} \div 2 = \frac{\square}{\square}(\ )\frac{\square}{\square} = \frac{\square}{\square}$$

They complete it in 23 seconds (19:43–20:06). In eight more seconds (20:10–20:18), the class call out their answers as part of the whole class interaction prompted by the teacher.

(2e)   Posing problems, looking for patterns and reconsolidating the method

Transitioning to the next step, Ms Q asks the class in what other way 4/5 can be equally divided in addition to being halved as in the first task. The class call out 4, which is affirmed by the teacher. She then prompts whether it could be equally divided into five sections. With the class agreeing to her, she writes $\frac{4}{5} \div 5$ on the

a

Lesson Structure

■ 1. Activating prior knoweldge   □ 2. Developing new knowledge   ▨ 3. Consolidation and formative assessment   □ 4. Summary and sublimation

Part 1 (00:00–02:36)

b. Task Series #1

c. Task Series #2

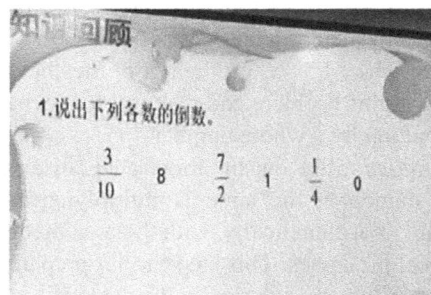

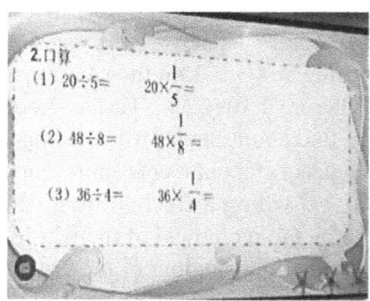

Part 2 (02:36–27:08)

d. Task #3

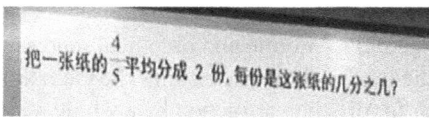

[Translation] If 4/5 of a piece of paper is divided into two equal parts, what fraction of the paper does each part represent?

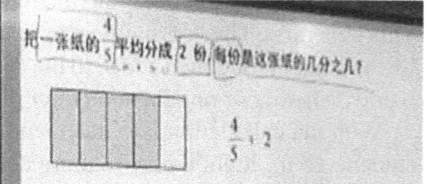

FIGURE 2.10   Ms Q's lesson delivered at the teaching research group meeting

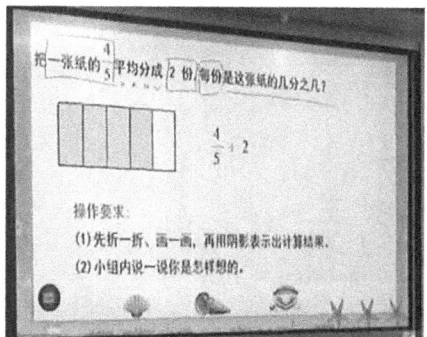

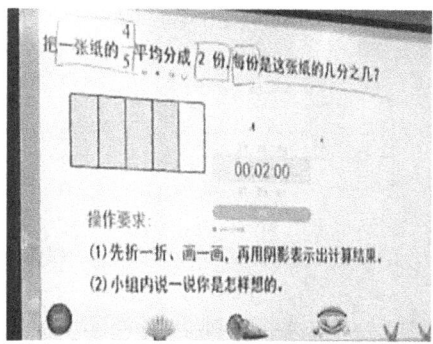

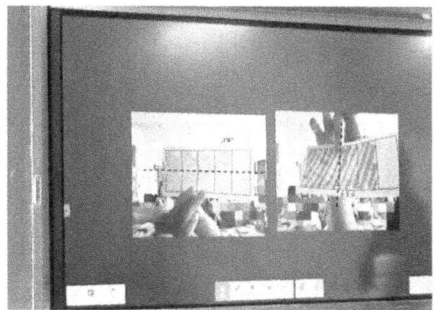

Note. The two black dotted lines were added by authors to enhance the folding lines that the two students made to divide 4/5 evenly into two halves.

Student work sample 1 (method 1)

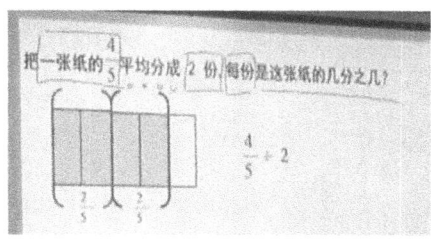

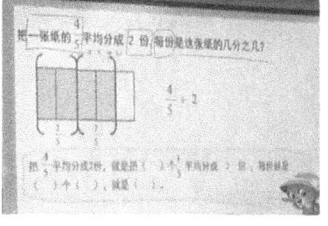

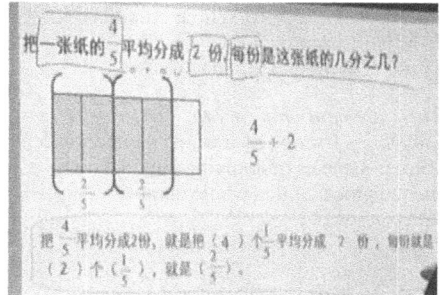

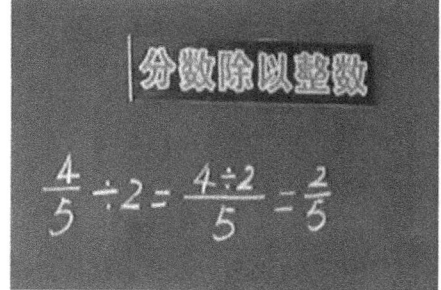

FIGURE 2.10 (Continued)

Student work sample 2 (method 2)

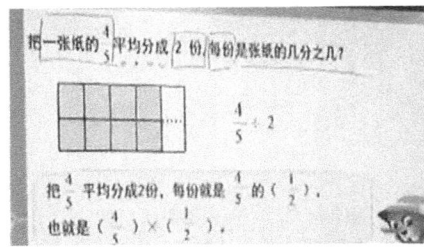
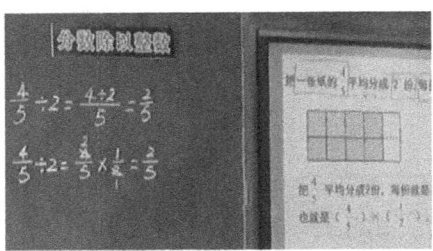

e. The more generalisable method (Task Series 4 and 5)

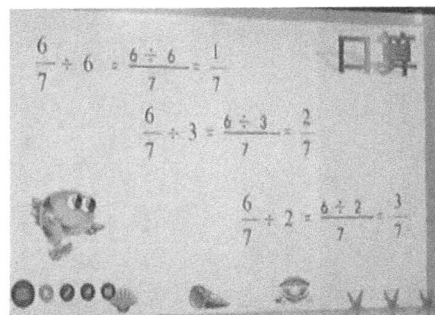
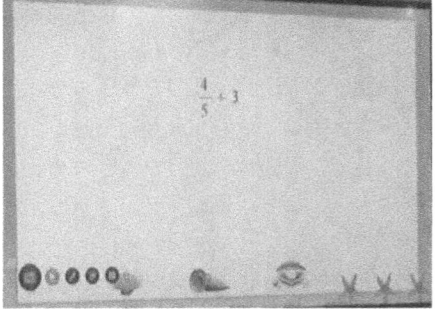

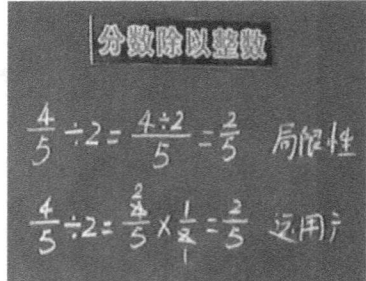

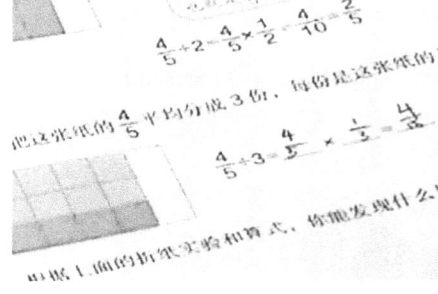

Translation:
分数除以整数: Dividing a Fraction
                  by a Whole Number
局限性: Limitated
运用广: Widely applied

The statement underneath:
Dividing a fraction by a whole number (except zero) is same as multiplying the fraction by the reciprocal of this whole number.

FIGURE 2.10 (Continued)

如果把这张纸的 $\frac{4}{5}$ 平均分成 3 份，每份是这张纸的几分之几?

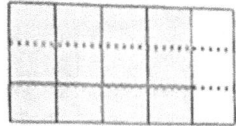

$$\frac{4}{5} \div 3 = \frac{4}{5} \times \frac{1}{3} = \frac{4}{15}$$

Task series 6: speedy exercise in 23 seconds (19:43–20:06)

$$\frac{9}{10} \div 3 = \frac{\square}{\square} \times \frac{\square}{\square} = \frac{\square}{\square}$$

$$\frac{3}{8} \div 2 = \frac{\square}{\square} (\boxdot) \frac{\square}{\square} = \frac{\square}{\square}$$

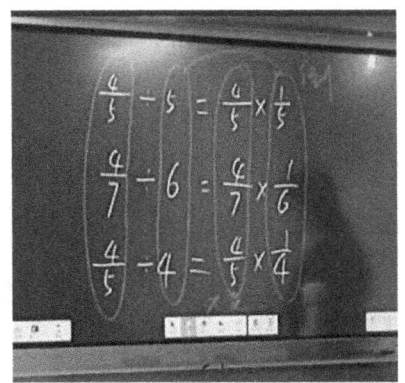

f. Task series 7–8: summarising and consolidating the method

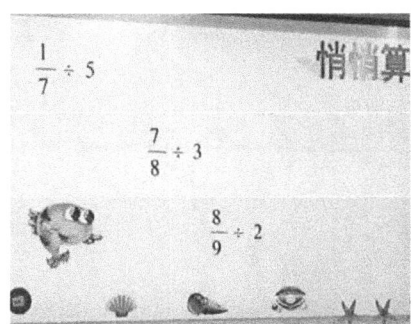

分 数 除 法 要 注 意
一 个 不 变 两 个 变
被 除 数 不 用 变
除 号 变 乘 号
除 数 变 倒 数

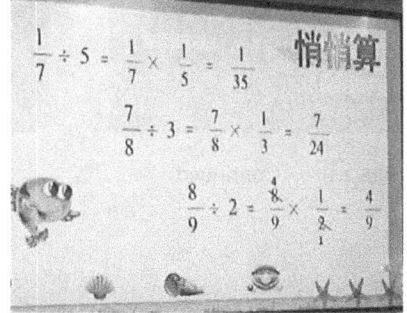

FIGURE 2.10 (Continued)

Part 3 (27:08–39:09)

g. Task series 9–12: speed, accuracy and error-proofing

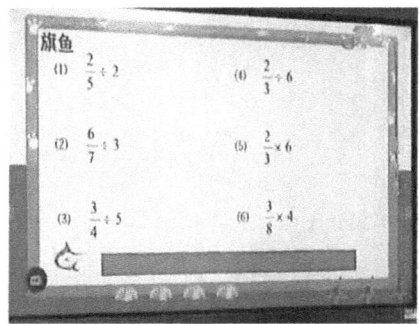

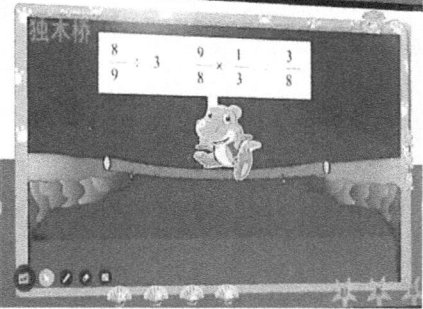

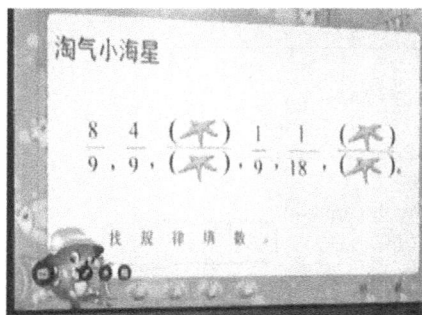

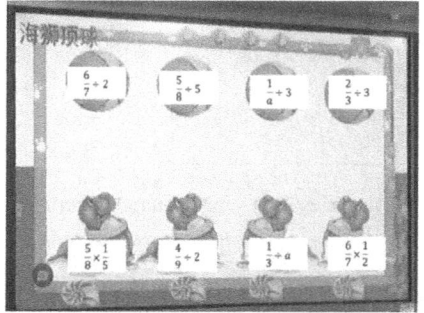

Part 4 (39:09–41:02)

h.

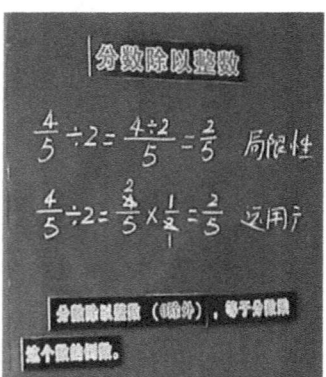

i.

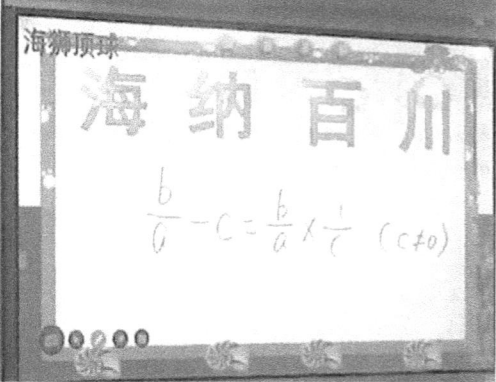

FIGURE 2.10   (Continued)

smartboard and turns to the class, asking, 'Apart from paper, what else could be divided into five equal sections?' The class then start to pose real-world problems for $\frac{4}{5} \div 5$ in addition to the paper situation. Four volunteers are invited. The first gives a cake situation. The second an apple. The third a bar of chocolate. The fourth says, 'Divide one-fifth of four cakes into five even sections and take one section.' The teacher looks excited, 'You've given this deeper thought.' She concludes this task by acknowledging that that are many examples in the real world.

This task seems to carry three connecting roles: (1) transitioning naturally from the previous content in this lesson to this step (connections between tasks); (2) offering opportunities to think about wider real-world situations (connections between mathematics and rich contexts); (3) preparing the class for the next step (connections between tasks). All in all, using this task makes the lesson more coherent.

By applying the newly established calculation law, the whole class call out their solution processes to two more tasks posed by Ms Q, 4/7 ÷ 6 and 4/5 ÷ 4. Then, the teacher asks two pivotal questions, pointing from the three tasks (4/5 ÷ 5 and these two tasks) to their solution processes on the board, 'Look carefully. What did not change? What changed?' The subsequent dialogue leads to three conclusions: (1) the dividends have not changed; (2) the divisors have turned into their reciprocals; (3) the operation has shifted from division to multiplication.

These conclusions are further summarised in the following slide as a rhyme which is a feature of Ms Q's teaching. It helps the children remember something they have understood but not necessarily absorbed entirely in one go. The rhyme reads, '分数除法要注意，一个不变两个变，除号变乘号，除数变倒数'， meaning: pay attention to fraction division where one thing does not change; two things change; division sign turns into multiplication sign; divisor becomes reciprocal. Children are asked to recite the rhyme twice: once in pairs and once together with everyone in the class. Such speedy reciting is based on thorough understanding that has been cultivated before it.

(2f) Consolidating the knowledge with three more tasks

Immediately after that, they are asked to tell their desk-mates the solution processes and results of three new tasks on the slide: $\frac{1}{7} \div 5$, $\frac{7}{8} \div 3$, and $\frac{8}{9} \div 2$.

Following this are two questions from the teacher, 'What have we just learnt? How do we solve it?' The class call out their answers to these questions. Answering the two questions helps further consolidate the newly learnt knowledge in children's mind. These tasks also work as a transition point for the lesson to move from Section 2 to the following section where the class work together to solve a series of tasks in various forms, with the teacher formatively assessing students' understanding of the focal knowledge.

2.10.3   Part 3 of 4: Consolidating and formatively assessing knowledge (27:09–39:09)

Demanding speedy mental maths, the first six tasks are opened by a sailfish. Children get their pens and worksheets ready before Ms Q announces, 'Get set, ready, go!' Then the class keep their heads down and quickly write the solution processes and answers, with the digital clock making a tick-tock sound in the background.

Then, the second task is themed as a game where the sailfish is allowed to pass a bridge by giving a correct statement on the topic learnt today, *Dividing a Fraction by a Whole Number*. If the statement is incorrect, it will fall off the bridge. There are animations and sound effects after each subtask.

The teacher seems excited to find errors emerging from the class, as if she has discovered a magic wand that can transform learning. Demanding speedy response from the class, the true or false tasks in this section create more than fun and rhythm. There are certainly deeper purposes embedded underneath. In case of no errors in the class, Ms Q has prepared typical errors or tasks that are easy to get wrong and she keeps them, waiting for the children to experience before the lesson ends. To master the knowledge, everyone has to encounter and combat mistakes, one way or another. Misconceptions thus play an important role in the lesson and, in fact, in any lesson.

The third task is called Naughty Starfish which covers the numerator and denominator of two fractions in a sequence of six fractions. The children need to figure out the pattern of the sequence and then the numbers hidden beneath the four starfish who are waving their arms adorably. From the PowerPoint, a female voice speaks up, 'Think about it. Can you solve it?' Ms Q says, 'That's my voice 18 years ago.' Children seem astonished and delighted, exclaiming: 'Wow!'

The last task is called Seals Hitting the Ball. On the slide, there are four balls each corresponding to one of four seals below – on the 'ocean floor'. A number sentence is stuck on each ball and every seal. This is in fact a matching problem consisting of four pairs of number sentences that have identical answers. Each time when a student matches a pair correctly and justifies the answer with thorough explanations, the teacher touches the screen and the seal rises to hit the ball. Then, both the seal and the ball disappear, unveiling one Chinese character hidden behind the ball.

In the end, a four-character Chinese idiom appears which reads '海纳百川', literally meaning 'all rivers run into the sea', and which metaphorically means 'one should have a broad mind'.

Ms Q says, 'I hope all of you have knowledge as wide as the sea and mind as broad as the ocean.' The tasks now go beyond mathematics and contribute to the affective development of learners.

2.10.4   Part 4 of 4: Summarising the new knowledge (39:10–41:02)

Clicking on the pen icon on the board, Ms Q takes a few steps forward and asks, 'So, what have we learned in today's lesson?' The class respond immediately, 'Dividing a fraction by a whole number.'

(4a) Summarising the knowledge verbally

Stepping backwards to the left chalkboard and pointing briefly at the title and content there, Ms Q bounces another question back to the class, 'Dividing a fraction by a whole number. Who would like to tell us the calculation method?'

With many putting their hands up, Ms Q invites a girl at the back to answer.

The girl says, 'Dividing a fraction by a whole number, except for zero, is the same as dividing the fraction by the whole number's reciprocal.'

Ms Q asks the class, 'Right?'

The class respond, 'Yes!'

Ms Q then asks the class to tell each other quietly the calculation method one more time. Immediately, everyone is murmuring to the person next to them, 'Dividing a fraction by a whole number, except for zero, is the same as dividing the fraction by the whole number's reciprocal.'

(4b) Summarising the knowledge algebraically

Then, Ms Q asks a final key question, 'How do we represent the solution method with letters?'

The class start to murmur amongst themselves, with Ms Q smiling and looking around the class. Before the lesson ends, Ms Q and the class state the equation together, with the teacher jotting it down on the smartboard (Figure 2.10i).

The teacher and class bow and say goodbye to each other.

2.11   Meeting on the spot after the lesson

The teaching research team come forward from the rear of the class, pick a student seat and sit down. Ms Q initiates the teaching research session by restating the title of the lesson. She then reflects on the elements that she planned but decided to drop. Looking back, the lesson features helping students to understand two methods of calculation through the combination of symbolic and graphic representations (数形结合). She says that children can understand and have a sense of the rationale but not sufficiently (a common way of being humble in the Chinese culture).

Then the head of the Teaching Affairs Office, Mr L, who also participates in the study as one of the local master teachers, shares his view about the lesson. He mainly talks about the strengths: (1) the ICT skills is really impressive, for which he proposes a themed TRG meeting the following week so Ms Q could give the group a tutorial on how to use the smartboard; (2) the lesson is based on prior knowledge, fraction multiplication and reciprocals, which is well covered by the warm-up exercises; (3) all the exercises are well arranged with the difficulty level gradually increased; (4) the hands-on activity is a highlight; (5) the atmosphere is really good.

He says, 'I could never teach a lesson on a usual day like this up to such quality.' Ms Q looks very humbled. He talks about two aspects that could be better. First, the hands-on activity could have allowed more opportunities for children to talk about their understanding. Second, in the last bit of the lesson, the algebraic expression,

$\frac{b}{a} \div c$, should have one more condition in addition to the condition $c \neq 0$, that is $a \neq 0$. Other colleagues join in and most of them think adding it is unnecessary, since $\frac{b}{a}$ is already given as a fraction and the denominator of any fraction should be nonzero. After the exchange of ideas between Mr L and colleagues, Ms Q acknowledges that his suggestion is indeed quite right and necessary.

Noticing that everybody has stopped talking, the municipal teaching research official, a former maths teacher now near retirement, speaks up. He emphasises that the lesson has a very warm atmosphere and is supported by very good ICT skills from the teacher, which is needless to say. What he most wants to share are the following few points. Rationale for calculations is one of the hardest types of content to teach. This particular lesson provides the base for the next lesson which is *Dividing a Fraction by a Fraction*. Lessons on calculation have two common characteristics: (1) it requires full understanding; (2) it relies strongly on prior knowledge. Due to these two characteristics, it is very important to activate the relevant prior knowledge that is closely related to the new. For example, this lesson relies heavily on Multiplication of Fractions. He acknowledges in this lesson:

> children had the opportunity to draw, calculate and then think deeply about the rationale behind the calculation. Having got the result, they then looked back to think about the reason. . .. The arrangement of the lesson content was well thought out, and the procedure was connected to the results which in turn led to the discussion of the rationale behind the calculation. However, the reason for using reciprocals had not been thoroughly discussed. The discussion could have been fairly straightforward: because 4/5 had been split evenly into two sections, one of which would be ½ of it [i.e., 4/5]. With this discussion, the understanding of the calculation rationale could have been deepened.
>
> (*MasterMT, 20180928, the municipal teaching research official's comments on Ms Q's lesson*)

In his view, thinking about the essence of mathematics is important, since the essence of mathematics is generalisable. He affirms that the teacher did quite well in this regard by arranging new problems after exercises. It is also commendable that the teacher not only expanded children's knowledge in mathematics but also beyond mathematics. For this, he points out a typical example – Ms Q set the lesson in an oceanic situation and got the children to know a new type of fish, sailfish, one of the speediest swimmers in the ocean. He continues, acknowledging that the lesson consists of thoughtful designs in a seemingly casual layout, with exercises applaudably set in an oceanic context. From his perspective, two aspects could have been done better. First, he thinks that more consideration could have been given to the entire class. When inviting children to talk about their solution processes, more time could have been allocated so that more students had more opportunities to talk. Second, it is questionable whether the calculation rationale should be summarised in exact words when the main thing here was to 'understand' the rationale.

Other teachers are shy to speak up due possibly to the power distance often identified in the Confucius heritage culture. They are more natural in the second TRG meeting presented next where no external 'leaders' are present. However, on this occasion, as part of the TRG routine, they all made thoughtful notes (Figure 2.11). There are internal and covert conversations between them and the senior and more knowledgeable others who have spoken up. The meeting is closed by Mr L who announces that next Friday during the usual teaching research slot Ms Q will be teaching everyone how to use the smartboard.

FIGURE 2.11   Observation notes written by teachers during the TRG observation session

This TRG meeting has the typical structure of Chinese lesson study. Perhaps the only missing piece is the planning stage following which are: (1) implementing (teacher) and observing (colleagues) the lesson, (2) reflecting on the lesson implementation, and (3) finding points of improvement. With these in mind, now we step into the second TRG meeting and see the natural leading position that a master teacher is put in as she shares her knowledge and experiences with peers.

2.12   TRG meeting 2 of 2

In the TRG meeting on 12 October 2018, Ms Q gives a presentation on how to use the newly equipped smartboard in the school. This time the meeting involves slightly more than the maths department. Sitting in the school hall are around 20 colleagues from across the school teaching major subjects, including mathematics. Five of them are student teachers from the partnership university who are here for school placement. The deputy head teacher attends the meeting, too. Snacks are provided. Everyone is happy, talking to each other before the demonstration starts.

Although many schools in the remote countryside of the province are already using a smartboard as part of the provincewide policy in supporting rural education, Ms Q's school has only managed to catch up with the trend over the summer break this year (2018). Ms Q says she learnt to master the smartboard skills in her prior position in an independent school ten years ago. Ever since joining the school, she has been looking forward to the school updating their IT facilities, and now it is happening.

Ms Q starts her presentation with a cover slide titled 'Brightening Up the Classroom Teaching with Multimedia'. Next she starts with an introduction to the software on the market from the past to the present, drawing on her own experiences.

Looking back, she says in an almost joking tone that she set off on her ICT-learning journey as a new teacher in an independent school in 2000, not knowing how to start a PC, when what she desperately wanted to learn was how to use PowerPoint. Many other teachers at the school then could already do loads of amazing things on a computer. It was jaw-dropping for her to see animations they made. 'I was like: wow, that apple can move!' With a humble circumstance back then, owning a computer at her rented home was impossible. The only option was to learn to use the computer in the school over the weekend. She was deeply interested in PowerPoint the first moment she had a blank file opened up in front of her. Gradually, she learnt to use the then popular software Authorware to integrate with PowerPoint but later found the file size often ended up being huge, so she taught herself to use Flash. A colleague in the audience nodded her head, whilst jotting down notes. Since the age of 18 when she graduated from the normal school (中等师范学校), Ms Q has had no opportunities to be taught to use the new technology.

'I have to teach myself everything. There's no other way around this,' she says.

Many teachers of Ms Q's age in this country trained as a teacher in a normal school which falls into the upper-secondary stage catering for students aged

15–18 years. They had been found to be mathematically and pedagogically more competent than their international counterparts holding bachelors and/or masters degrees (Ma, 1999). Most of them, including Ms Q, by now should have at least earned a bachelor's degree by passing specific exams for independent study, but the foundation for teaching the subject was laid back in their normal school days.

'Nowadays,' she says, 'PowerPoint has almost replaced Flash and can perform all sorts of animations and effects.'

She continues, talking about the art of PowerPoint and her tips to PowerPoint design. She reminds colleagues that the design should first start from the grand plan. The key is the teacher *not* the tools, and the purpose of using PowerPoint should be improving teaching.

Then she emphasises that all slides should be simple. The meaning of PowerPoint is literally the *key point*. There should be fewer words and more graphs. PowerPoint should serve the audience. Each slide should not have more than six lines. The slide should be of great contrast. The font size should be big enough. There then appears a slide with the same words of different sizes which Ms Q uses to show contrasts between font sizes (Figure 2.12a). By chatting back and forth with the colleagues, she says, '40 is indeed too big, and 34 is about right.' Then, in the following slide, she quotes the four by six rule: no more than six lines of text per slide and six words per line, visible from six pigeon steps, and understandable in six seconds.

Ms Q summarises two common features of bad PowerPoint design: filling slides with words and using PowerPoint the 'Word' way. With the quote 'a good picture is worth a thousand words' shown on the screen, she concludes this section of presentation on the importance of using visual in PowerPoint.

At this point, she reminds colleagues that though the smartboard has PowerPoint, it should not only be used that way or as an average whiteboard. She then proceeds to show a wide range of functions that the smartboard can perform in teaching mathematics and many other subjects, giving a detailed demonstration about how to (Figure 2.12):

- do handwriting;
- add pages;
- use a ruler to draw ruled lines so as to demonstrate in the class the standard way of writing specific solution processes in a notebook;
- erase part of or the entire page;
- draw various shapes, such as a circle or star, and make duplicates;
- use the magnifier app;
- use the board-in-board function;
- take a screenshot with the smartboard;
- use the timer app that comes with the board;
- use the spotlight app.

FIGURE 2.12 Master teacher Ms Q showcasing the use of smartboard in a TRG meeting

p-s

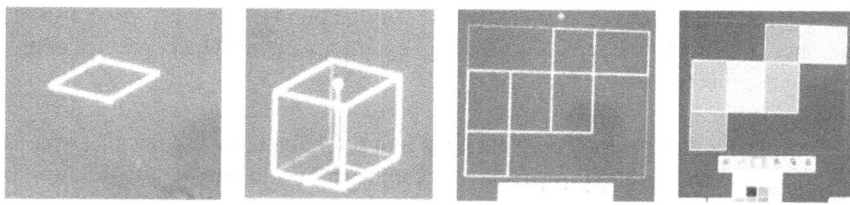

t-v

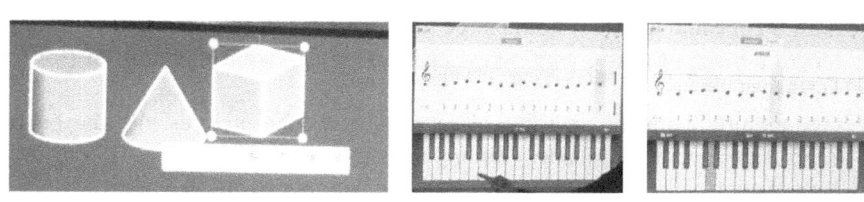

w-x

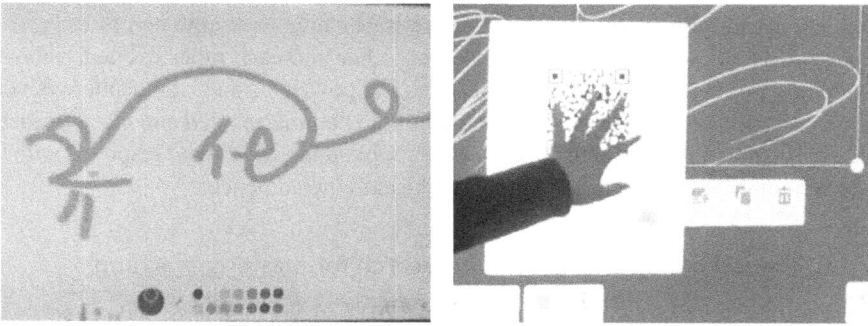

FIGURE 2.12 (Continued)

Then she shares a few tips for colleagues who teach Chinese. After a few quick clicks on the board, several Chinese characters appear stroke by stroke. Several more clicks on the board lead to the appearance of their *pinyin* (Chinese phonics) in a row with four tones (Figure 2.12c–d). In just a few seconds, teachers understand how to use the smartboard to teach the mother tongue, Chinese, where strokes and pronunciation of characters are fundamental. These will be particularly useful for the lower grades. Noticing that Ms Q is demonstrating the special functions of the smartboard for almost all subjects, a colleague in the audience excitedly announces: 'We should have the entire teaching staff here.'

After giving an introduction to the use of the smartboard in a variety of subjects, Ms Q goes back to the mathematical tools that can be fairly smart for classroom teaching (Figure 2.12g–t). First, she draws a cylinder, showing the colleagues how to enlarge and shrink the shape by dragging the right bottom corner, how to unpack it into a net and how to colour the net. At various points, there are wows from the audience whilst Ms Q keeps revealing the exciting content. She continues demonstrating the methods to:

- draw a cone, unfold the cone into a net and set the bottom and the side of it into different colours, then fold the net back into a cone and drag it to the right of the cylinder on top of the slide, putting them on 'exhibition' as teachers would do about mathematical widgets that have just been covered in a lesson;
- draw a cube and unfold it in the three kinds of ways (e.g., the '141' format);
- record a piece of music with e-keyboard and then have it replay itself (for the music class);
- take pictures with the smartboard camera, insert the pictures into the board, adjust sizes and do duplicates (for any subject);
- remote control the board from a smartphone with which one can take pictures of students' seatwork and send them wirelessly to the board (for all subjects).

After showing how an arts teacher can use it, she announces the end of the presentation. Then, her colleagues start to chat to her and each other excitedly about how they might use it in their own classes. The power-distance atmosphere never appears in this session where everyone seems so natural, relaxed and close to each other. The meeting ends with the colleagues applauding Ms Q who responds with a big yet humble smile, gently nodding her head to the audience.

2.13   Embarking on the nationwide search for masterly teaching

Over the 40 days spreading across the year 2018 in Ms Q's class and school, we got to think about the wider project and how we should adapt our proposal to the reality. Over the course, we got access to the entire community of primary mathematics teachers in the province and beyond. We are now ready for the nationwide search and exploration into masterly teaching and learning. The following chapter presents the ways through which the project is carried out across five Chinese provinces – Anhui, Beijing, Jiangsu, Jiangxi and Tianjin.

3
GETTING INTO MASTER MATHEMATICS TEACHERS' CLASSROOMS

Chapter overview

- The Integrated Paradigm
- The MasterMT framework
- Seeing a world in a grain of sand and the masterly teaching in a case study
- Looking for the master teachers and their students
- Measuring and understanding teaching

  - Filming a usual maths lesson on a usual school day like a member of the class
  - Capturing the lesson through both the teacher's and the learner's eyes
  - Constructs for quantitative observation
  - Qualitative observation

- Measuring and understanding learning

  - Collecting sociodemographic information
  - Measuring affective learning outcomes in mathematics
  - Measuring metacognitive learning outcomes in mathematics
  - Measuring cognitive learning outcomes in mathematics

- Modelling the teaching-learning mechanisms
- Studying the development of master mathematics teachers
- Multiple perspectives, multiple layers
- Multiple evidence, one reality: Ready for the MasterMT journey

> A journey of a thousand miles begins with a single step.
> – Laozi (400 BC, Tao Te Ching, Chapter 64)

DOI: 10.4324/9781003127925-3

After the in-depth case study, we are ready for a large-scale exploration into mathematics teaching and learning in more master teachers' classrooms. Our ultimate goal is to capture best mathematics teaching practices and mechanisms behind masterly teaching and learning in primary schools. It is approached through three lines of enquiry: (1) quantitative measurements of learning and teaching in multi-dimensions as well as qualitative interpretation of teaching and learning in multiple layers; (2) multilevel modelling of teaching effects on and (multilevel) structural equation modelling of paths towards multiple learning outcomes; (3) deep listening and investigation into teachers' accounts of the mathematics lessons just delivered, their teaching beliefs in general, their PD trajectories towards masterly teaching – career stories told by themselves, and teacher PD in action – researchers' participatory observation of teaching research events at the school, municipal, provincial and national levels.

This chapter illustrates the project's research methodology. The remainder of the chapter starts with the clarification of the overarching research paradigm, Integrated Paradigm, and the theoretical frameworks of the project operated at macro and micro levels. We then look at the case study design as part of the project and the sampling criteria, procedures and results. After that, we explain in detail how multiple methods are utilised to collect and analyse the bulk of multicategory data. The chapter concludes upon the ways in which we synthesise the multilevel and multi-layer findings across multiple methods within the Integrated Paradigm.

3.1   The Integrated Paradigm

The MasterMT project is the second voyage of the Integrated Paradigm which came into being in the process of the authors' former study on effective mathematics teaching (EMT) in England and China (Miao & Reynolds, 2018; Miao et al., 2021). A set of quantitative (QUAN) and qualitative (QUAL) methods were carefully selected to serve the research purposes, with an equal emphasis given to deeper interconnections of initial results and findings of all methods. Despite this, the solution is not to simply mix two approaches together, with one serving the other. In the end, it is not a question of QUAN or QUAL or as simple as vaguely mixing both to a varying degree. It is about collecting relevant evidence from different angles and putting different pieces of evidence back together to represent the focal reality as holistically as possible. This is the underpinning feature of the Integrated Paradigm where major paradigmatic conflicts cease to exist.

At the heart of the Integrated Paradigm are three key features: multiple measurements, multiple perspectives, and integration of multiple types of evidence in multiple layers at multiple levels.

Multiple measurements. The Integrated Paradigm measures the focal issues, for example teaching and learning in the MasterMT, from different angles to see various facets of them. Learning is measured in three major dimensions: (1) metacognition (observation and questionnaire), (2) mathematical reasoning (observation and maths test) and (3) affects towards mathematics, maths teacher and the school (observation and questionnaire). Teaching is also measured in multiple dimensions: (1) behaviour (observation); (2) general teaching knowledge (observation and interview); (3) maths knowledge (observation, questionnaire and interview); (4) maths pedagogical content knowledge (observation, questionnaire and interview); (5) teacher metacognition (observation and interview); (6) teacher beliefs (observation, interview and questionnaire).

Multiple perspectives. The Integrated Paradigm cherishes multiple 'voices' from different roles by actively engaging them in the conversation on the focal issues. This allows the outsiders to spot the 'elephant in the room' that might be otherwise unseen by the insiders whilst offering insiders' opportunities to tell their side of story, the story behind the scene, and/or the beliefs and thinking hidden beneath the visible and therefore more obvious doing. In fact, insiders and outsiders are relative roles. Both participating teachers and researchers are insiders in the maths teaching profession and outsiders to each other's practice; researchers are outsiders to practitioners and their colleagues; colleagues from the same school are outsiders to each other in the individual sense; two international researchers/co-authors of this book are outsiders to the Chinese education system. Views from different roles form multiple perspectives around maths teaching and learning. These roles, often paired, also include learners and teachers. Efforts devoted to seeing things from multiple perspectives generate critical and balanced views from multiple sides, such as the researched vs. the researching and the Chinese vs. non-Chinese academics, which contributes to a more holistic understanding of the reality.

Integrating multiple strands and layers of evidence at multiple levels. The final stage of the Integrated Paradigm focuses on systematic integration of multiple strands and layers of evidence at multiple levels. This features the syntheses of results and findings from various methods, weaving together wisdom from QUAN and QUAN, QUAN and QUAL, and QUAL and QUAL, where various methods relate to each other via underlying meanings, free from artificial superficial paradigmatic barriers.

3.2 The MasterMT framework

The contribution of the MasterMT project to the evolution of the Integrated Paradigm is the development and utilisation of a framework with dual parts that collectively represent the reality at macro and micro levels (Miao et al., 2021). The macro part of the framework (Figure 3.1) is similar to the EMT framework; the micro part of the framework (Figure 3.2) zooms in to show detailed elements of teaching and learning, that are systematically looked into, and the interconnections within and between them.

The MasterMT framework further inherits the merits of both mathematics education research and teaching effectiveness research in that it applies both statistical analyses and modelling – this time more advanced – and in-depth interpretations of

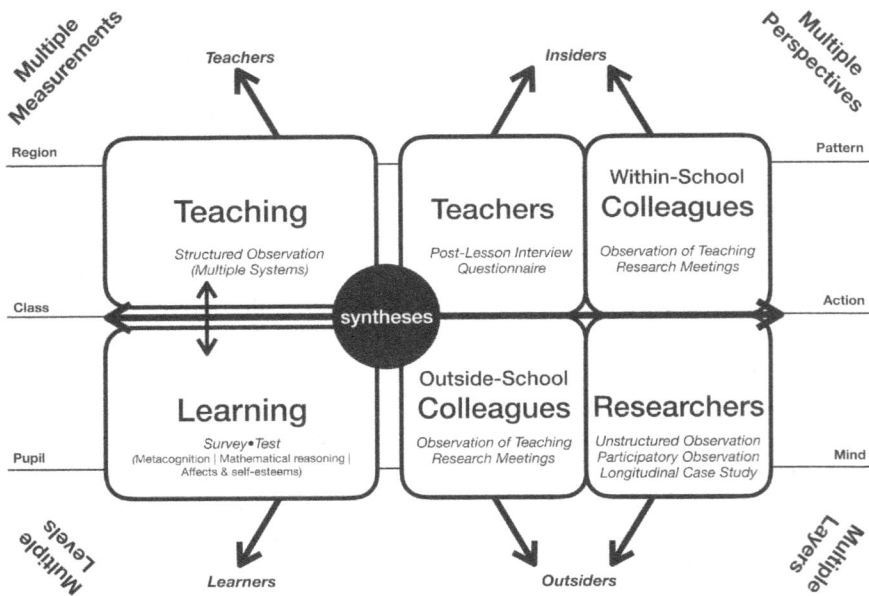

FIGURE 3.1 The MasterMT framework (macro)

multiple voices and details, including deep 'listening' of multiple voices from key roles in the focal field and detailed 'sketching' of mathematics itself and mathematics in the mind of 'mathematicians', young (the learners) and mature (the teacher). The final touch of the framework lays on interweaving multiple strands of evidence together to represent and explain the holistic reality.

In the MasterMT project, teaching is measured with a hybrid of multiple observation systems looking at different aspects of teaching and learning processes; learning is measured with multiple constructs focusing on multidimensional outcomes of learning. Teachers, their colleagues and the researchers offer their views on master maths teachers' teaching. The ultimate integration of evidence sees multiple statistical analyses, such as multilevel structural equation modelling, joining the in-depth analyses of qualitative data, such as a case study of a master teacher's work over a period of 40 school days (initially planned for a month) and participatory observations of teaching research meetings/conferences at the school, municipal, provincial and national levels.

3.3 Seeing a world in a grain of sand and the masterly teaching in a case study

To have an in-depth participatory observation of a master teacher's work, the first author spent about two half-semesters (40 school days) from April to October 2018 in a local primary school where the master mathematics teacher, Ms Q, works. The

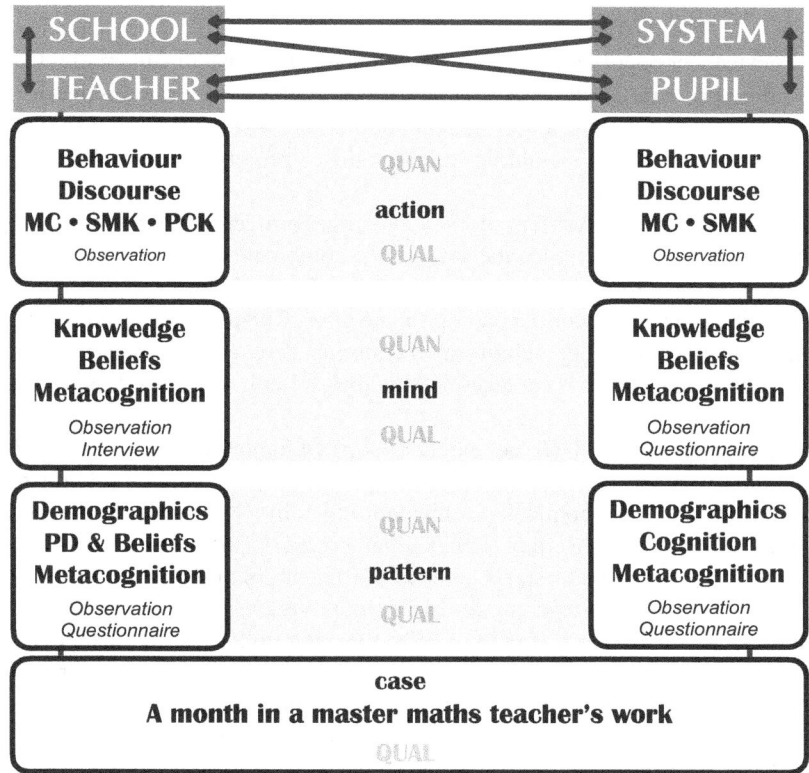

FIGURE 3.2 The MasterMT framework (micro)

Note: SMK = subject matter knowledge. PCK = pedagogical content knowledge. PD = professional development. MC = metacognition.

school also offers the project precious opportunities to pilot some of its methods. Besides observation of daily maths lessons, most of the days we follow the teacher through half of her daily work routines (8 am to 12 pm). We are thus able to look closely at: her maths teaching over 40 school days; her work in an open-space office shared with five colleagues, from lesson preparation and reflection to homework grading, from communication with students and colleagues to recording and analysing student progress; weekly teaching research group meetings; supervising trainee teachers and mentoring visiting teachers from across the province.

The process brings us closer to the masterly teaching reality than just relying on the review of research literature alone. With the rich information accumulated over the course of the case study, we are also able to further extend our networks with the practitioners' sector in more meaningful directions and compile an optimal plan for collecting the project's major data in more schools and provinces.

3.4 Looking for the master teachers and their students

The target teacher population of the MasterMT project is master mathematics teachers teaching primary Grades 5–6 in capital cities of provinces or urban districts of municipalities in China; the target student population is the students in these teachers' classes. The teachers should be well recognised provincially or nationally for excellent maths teaching.

Piloted in Spring and Autumn of 2018, the larger project recruited master mathematics teachers and their students, through the combination of stratified sampling and expert recommendation which located cities first and then teachers and their classes, with one teacher corresponding to one class. The project was approved by the first author's institution, and informed consents were obtained before data collection. Three major steps were taken in obtaining a final sample for the project:

- First, according to the GDP per capita ranking of major cities in China in 2018 (National Bureau of Statistics of China, 2019), we located the sampling sites – capital cities and economically equivalent cities in five Chinese provinces or municipalities, Anhui, Beijing, Jiangsu, Jiangxi, and Tianjin. These cities spread about evenly across the first, second and third quarters of the 2018 ranking.
- Second, the provincial teaching research officials responsible for primary mathematics, as gatekeepers of mathematics teaching and learning in their home provinces, were asked to help recruit around 15 master mathematics teachers who demonstrated best teaching practice in primary mathematics and were full-time classroom teachers teaching mathematics to primary grades, ideally Grades 5 and 6. The inclusion criteria seek to look for master teachers who demonstrate best primary mathematics teaching in their home province and: (1) have the ranks of Professoriate-Senior or Senior; (2) are recognised as Super Teachers, Subject Leaders or Backbone Teachers; (3) have won teaching awards in provincial and/or national teaching competitions.
- Based on the teaching research officials' recommendations, we have an initial sample of 81 teachers and 3,737 students. With informed consent from participants, we collected the initial bulk of data in the 81 teachers' classes, including lesson observations. To make sure that the teaching data collected represented daily teaching that would happen the same way without the appearance of external observers and cameras, we applied one more criterion (the third and last step) to screen and finalise the sample based on the authenticity of the lesson data, though every effort had been taken to minimise observer effects during the data collection. The criterion was simple: the lesson must be a typical normal lesson that was unfolded as usual after its precedent lesson. This resulted in 70 teachers' lessons meeting the criterion. We had to drop data related to 11 teachers and their students because their lessons were carefully prepared teaching-research lessons (教研课) which did not represent the kind of everyday teaching we were explicitly aiming to observe.

The final sample of the project (Table 3.1) included 3,178 students in Grades 2–6 and 70 master mathematics teachers of which 59 were teaching Grade 5 or 6 in the 2018–2019 academic year. Of all the teachers, 14 are from Anhui, 16 from Beijing, 15 from Jiangsu, 14 from Jiangxi, and 11 from Tianjin. Like the general fact of gender imbalance in primary school settings, about 2/3 of the teachers are female ($n = 47$), with about 1/3 of them being male ($n = 23$).

TABLE 3.1 Demographic information of master teachers and their students

	Students			Teachers		
	N	*Age*	*%*	*N*	*Age*	*%*
Province	*3033*	*11.3 (0.9)*	*100%*	*70*	*41.5 (4.4)*	*100%*
Anhui	673	11.0 (0.9)	22.2%	14	43.0 (5.6)	20.0%
Beijing	523	11.5 (0.5)	17.2%	16	39.0 (3.6)	22.9%
Jiangsu	655	11.2 (1.1)	21.6%	15	42.4 (3.5)	21.4%
Jiangxi	733	11.2 (1.2)	24.2%	14	41.4 (4.1)	20.0%
Tianjin	449	11.7 (0.6)	14.8%	11	42.8 (4.3)	15.7%
Grade	*3033*			*70*		
2	54	8.1 (0.4)	1.8%	2	38.5 (3.5)	2.9%
3	134	9.3 (0.4)	4.4%	2	45.5 (0.7)	2.9%
4	203	10.0 (0.6)	6.7%	5	45.0 (4.5)	7.1%
5	1318	11.1 (0.4)	43.5%	32	41.2 (4.5)	45.7%
6	1324	12.1 (0.4)	43.7%	29	41.3 (4.2)	41.4%
Gender	*3028*			*70*		
Female	1392	11.2 (1.0)	46.0%	47	41.3 (3.9)	67.1%
Male	1636	11.3 (0.9)	54.0%	23	42.0 (5.3)	32.9%
Teacher Position within Schools				*70*		
Teacher with no other obligations				22	42.1 (5.9)	31.40%
Head or Deputy Head of Maths TRG				32	40.3 (2.9)	45.70%
Headteacher or Deputy Headteacher				16	43.1 (4.1)	22.90%
Teacher Rank				*70*		
First Level (一级)				18	39.9 (3.7)	25.7%
Senior (高级)				43	41.8 (4.5)	61.4%
Professoriate-Senior (正高级)				9	43.7 (4.5)	12.9%
Honorary Titles				*70*		
Backbone Teacher (骨干教师)				17	39.0 (4.0)	24.3%
Subject Leader (学科带头人)				40	41.6 (3.4)	57.1%
Super Teacher (特级教师)				13	44.7 (5.6)	18.6%
Mathematics Teaching Experience				*70*		
10 to 19 years				20	37.1 (2.0)	28.6%
20 to 29 years				46	42.6 (2.8)	65.7%
30 to 38 years				4	51.5 (4.0)	5.7%

Note: TRG = teaching research group (教研组). The project's student sample consists of 3,178 students of which all were in the lessons observed, 3,033 completed a survey and provided their demographic information in the course of completing the survey, and 145 students did not complete the survey because of scheduling difficulties. The student information presented in the table comes from the 3,033 students.

All 70 teacher participants are full-time mathematics teachers in their thirties ($n = 24$), forties ($n = 43$) or fifties ($n = 3$). The majority ($n = 46$) of the teachers have 20 to 29 years' mathematics teaching experience; 20 teachers have 10 to 19 years' experience; 4 have 30 to 38 years' experience. In terms of professional ranks, 18 of the teachers are First Level, 43 are Senior, and nine are Professoriate-Senior. In terms of honorary titles, 17 are Backbone Teachers, 40 teachers are Subject Leaders, and 13 are Super Teachers. They have all won multiple teaching awards in teaching competitions at provincial and national levels. In addition to their teaching obligations, 48 of them are also (deputy) headteachers or (deputy) heads of the teaching research group for mathematics at their home schools.

From five provinces/municipalities, 70 master mathematics teachers have taken part in the project, with an average age of 41.5 years ($SD = 4.4$, $Min. = 32$, $Max. = 57$). As most teachers in state schools in China, all master teachers are full-time. Their mathematics teaching experiences average 21.9 years ($SD = 5.1$), ranging from 10 to 38 years. Of all the 70 teachers, 9 are Professoriate-Senior teachers, 43 are Senior teachers, and 18 First-Level teachers. The majority ($n = 64$) of teachers have a bachelor's degree, with four having an associate degree and 2 masters.

With the participants recruited, we are ready to collect and analyse data. To address the research aims/questions, we construct a system of multiple research methods using the Integrated Paradigm. Before the ultimate integration, we carry out multiple measurements of teaching and learning, model the teaching-learning mechanisms at multiple levels and capture multiple perspectives from multiple roles through multiple layers of educational reality.

3.5   Measuring and understanding teaching

Teaching is measured and interpreted in multiple areas: (1) behaviour (observation); (2) general teaching knowledge (observation and interview); (3) maths knowledge (observation and interview); (4) maths pedagogical content knowledge (observation and interview); (5) teacher metacognition (interview); (6) teacher beliefs (observation, interview and questionnaire).

3.5.1   Filming a usual maths lesson on a usual school day like a member of the class

What we want to see is the way each master teacher teaches on a daily basis – the idea of 'home-teaching'. We do not want to see a carefully prepared lesson for outsiders to be impressed. The teacher should teach the lesson as she/he would even if there were no external observers. Measures are taken to reach this ideal quality of observation. Communication with the teacher is made ahead of observation about the research purposes and what their 'real' teaching means to the research findings which will be shared with them after completion of the study. During the data collection, every effort is made to minimise observer effects.

All teachers each have one lesson observed and video-recorded. As previously stated, amongst the participants, there is a case study teacher whose consecutive lessons are observed. Two cameras are utilised to capture the whole class and the teacher respectively. Both structured and unstructured lesson observations are carried out through the analyses of lesson videos. The following aspects are at the core of the lesson analyses: (1) teacher and student behaviours; (2) classroom discourse; (3) cognition and metacognition (MC) in action; (4) subject matter knowledge (SMK) in action; (5) PCK in action.

3.5.2 Capturing the lesson through both the teacher's and the learner's eyes

The placement of cameras is adapted from TIMSS Video Studies (Jacobs et al., 2003). Two cameras are utilised, with one being placed at the front to see the whole class from the teacher's perspective and the other at about halfway from the front of the classroom to capture the lesson from a student's view. Maximum effort has been given to placing both cameras alongside the window wall, so that the shot aligns with the direction of natural light. This is realised in most classrooms.

Structured observations are conducted using a hybrid of observation systems: (1) OTL (Reynolds et al., 2002), (2) ISTOF (Teddlie et al., 2006) and (3) MQI (Hill & Ball, 2004; Learning Mathematics for Teaching, 2011) capturing teacher and student behaviours, classroom discourse, cognition/metacognition in action, as well as teacher SMK and PCK in action.

The *Opportunity to Learn* (OTL) system was adapted from the International School Effectiveness Research project by Reynolds et al. (2002). This adapted version had formerly been utilised on the EMT project with lesson data collected from English and Chinese primary schools (Miao & Reynolds, 2018). The system captures five types of classroom activities and student time on task. In total, the OTL consists of six measures in the form of percentages.

For the quality of teaching, we utilised the adapted version of the *International System for Teacher Observation and Feedback* (ISTOF) (Muijs et al., 2018; Teddlie et al., 2006) as applied in the EMT project (Miao & Reynolds, 2018). The project dropped one of the seven dimensions, Differentiation and Inclusion, due to the fact that Chinese teachers took a whole-class approach in differentiating teaching which was not easily observable but generated much smaller performance differences amongst students than their English counterparts. The included six dimensions are shown in Table 3.2.

Measuring teacher knowledge for mathematics teaching in action, the *Mathematical Quality of Instruction* (MQI) was developed by Hill and Ball (2004) and colleagues on the Learning Mathematics for Teaching (2011) project. The MQI looks at teacher knowledge of both the content and students (Hill et al., 2008). Teacher knowledge for teaching mathematics has been a focus of theoretical and empirical endeavours for at least four decades (Ball, 1988; Ball et al.,

TABLE 3.2 Observation protocols for measuring the quantity and quality of maths teaching and learning

OTL	MQI (whole-lesson)
• Whole class interaction (OTL1) • Whole class lecture (OTL2) • Individual/group work (OTL3) • Classroom management (OTL4) • Partial class interaction (OTL5) • Student time on task (OTL6)	• Lesson time is used effectively (MQ1) • Lesson is mathematically dense (MQ2) • Students are engaged (MQ3) • Lesson contains rich mathematics (MQ4) • Teacher attends to and remediates student difficulty (MQ5) • Teacher uses student ideas (MQ6) • Mathematics is clear and not distorted (MQ7) • Tasks and activities develop mathematics (MQ8) • Lesson contains Common Core aligned student practices (MQ9) • Whole-lesson mathematical quality of instruction (MQ10)
ISTOF	
• Assessment and evaluation (ISTOF1) • Clarity of instruction (ISTOF3) • Instructional skills (ISTOF4) • Promoting active learning and developing metacognitive skills (ISTOF5) • Classroom climate (ISTOF6) • Classroom management (ISTOF7)	

2008; Hill & Ball, 2004; Kraft & Hill, 2020; Shulman, 1986). It is also at the core of teaching practice in China. The two vertices of the instructional triangle (Cohen et al., 2003; Lampert, 2001), the content and students, are one of the several aspects that Chinese teachers must talk about in a kind of PD activity taking shape in the 1980s, 说课 (pronounced as *shuo ke*), which literally means *Talking about the Lesson, Lesson Explanation*, or *Lesson Justification*. A lesson explanation typically happens before or after a teacher delivers a demonstration lesson to peers or external experts. It is essentially the teacher's justification of the lesson plan and how the plan went in a post-lesson meeting or conference. Conventionally, a teacher is expected to explain the lesson in terms of (1) curriculum and textbooks; (2) students and their characteristics; (3) teaching methods; (4) teaching steps; (5 teaching rationale, board notes and design.

The ISTOF inter-rater reliability was obtained between the first and the third authors on two lessons (translated) from China and two lessons from England on the Effectiveness of Mathematics Teaching project (Miao & Reynolds, 2018): $\kappa_1 = 0.78$, $\kappa_2 = 0.81$, $\kappa_3 = 0.84$, $\kappa_4 = 0.88$, $\bar{\kappa} = 0.83$. For the MQI whole-lesson measure, the first and the second authors reached agreement on two lessons with English subtitles: $\kappa_1 = 0.63$, $\kappa_2 = 0.78$, $\bar{\kappa} = 0.71$. The agreement was considered substantial by existing standards on Cohen's kappa (Landis & Koch, 1977).

Unstructured observations in the project allow variables to emerge from the data rather than from a predefined framework. We are conscious that there is always something missing from the existing observation systems combined. No existing systems can 'capture instruction in its entirety' (Charalambous & Praetorius, 2018, p. 357). Neither is there an appropriate system that serves well in analysing higher-order skills (such as metacognition) in action whilst catching the focal aspects that the project seeks in lessons more thoroughly. Analysing lessons without a set system, we will be able to look for, for example, the opportunity to learn two types of observable outcomes: (1) basic maths and (2) higher-order maths skills (mathematical reasoning and metacognition in action).

*Placement of camcorder*s. Following the methods suggested by the TIMSS Video Study (Jacobs et al., 2003), two camcorders were placed in the classroom beside windows where natural light comes in. One camcorder at the front pointing at the whole class, the other about a third of the way from the front follows the teacher, capturing the dynamics of the class.

Observation approaches. As introduced in Chapter 3, the lesson data were analysed both quantitatively and qualitatively. The quantitative analyses involved the use of three existing observation systems: the OTL, the International System for Teacher Observation and Feedback (ISTOF) and the MQI. The qualitative analyses drew upon multiple perspectives from teachers, their colleagues, the teaching research officials from local education authorities and the researchers.

3.5.3 Constructs for quantitative observation

The OTL measures the quantity of teaching in terms of proportions of lesson time allocated for whole-class interaction (OTL-1), whole-class lecture (OTL-2), student independent seatwork (OTL-3a) or groupwork (OTL-3b), class management (OTL4), partial class interaction (OTL-5), and student time on task (OTL-6). The current study added one subcategory into OTL3 – groupwork amongst more than two students (OTL-3c), categorising paired groupwork as OTL-3b.

The ISTOF measures the quality of teaching in a general sense originally in seven dimensions (Teddlie et al., 2006). Dimension one captures the quality of assessment that teachers use to diagnose students' understanding as they teach (ISTOF1). Dimension two considers teaching differentiation (ISTOF2), which was found to be not entirely observable in the Chinese context where differentiation was realised covertly (for detail, please see Miao & Reynolds, 2018, Ch4). As shown in Table 3.2, dimensions 3-7 focus on teaching clarity (ISTOF3) and skills (ISTOF4), the cultivation of metacognitive skills (ISTOF5) and class climate (ISTOF6) and management (ISTOF7).

The MQI measures the quality of mathematics teaching. To prioritise the data analysis schedule and acknowledge the fact that the segment- and lesson-level items of the MQI share similar underlying concepts, we decided to utilise the ten lesson-level items each rating the lesson on a five-point scale.

The explorative factor analyses (EFA) indicate an excellent internal reliability of the ISTOF as a whole ($\alpha = 0.97$), the MQI ($\alpha = 0.92$), ISTOF3 ($\alpha = 0.91$), ISTOF4 ($\alpha = 0.91$), ISTOF5 ($\alpha = 0.88$), ISTOF6 ($\alpha = 0.91$) and ISTOF7 ($\alpha = 0.90$), with the ISTOF1 ($\alpha = 0.68$) manifesting an acceptable reliability. After the EFA, to check the validity of the measures as latent variables, we performed a confirmatory factor analysis (CFA) on all ISTOF components that were utilised and the MQI. To examine the ISTOF as an overarching latent variable, we ran a second-order analysis where indicators (explained as follows) were umbrellaed by corresponding components which in turn were predicted by the ISTOF. Regarding the ISTOF indicators in the CFA, we utilised item parcels by averaging items within each indicator, since we were more interested in the indictors than individual items within them.

3.5.4 Qualitative observation

In complement with the quantitative analyses of lessons, we use the grounded theory approach (Cohen et al., 2018) to look for major features of teaching widely found across classrooms. In doing so, we do not carry any predefined system of observation. Rather, we allow the system/themes emerge from the lesson data. Typical lesson segments are selected as examples of each particular feature. The grounded theory approach endows the analyses of lessons with both depth and richness.

3.6 Measuring and understanding learning

Learning is measured in three major areas via a paper-and-pencil survey: (1) affects towards maths learning, maths teacher/teaching and the school (questionnaire), (2) metacognition (questionnaire), (3) cognitive outcomes in mathematics (test). For the formal data, students completed a booklet containing items measuring two (Grades 2–4) or three (Grades 5–6) of the focal domains, along with their sociodemographic information. The survey items, originally in English, were adapted for the project. They were translated, back translated, compared and field tested.

3.6.1 Collecting sociodemographic information

This part of the data mainly includes information about student gender, age in years and socioeconomic status (SES). They were deemed important variables for the understanding of maths performance variations across sociodemographic spectrums. Gender was coded dichotomously as 0 = boy, 1 = girl. Age was computed as a continuous variable in years by dividing the difference between the survey date and the child's date of birth by 365.25.

The SES is a latent variable inferred by three strands of information based on the classical SES framework: home possessions (SES1), parental highest level of education in years (SES2) and parental occupational status (SES3). Adapted from the TIMSS 2015 background survey items (IEA, 2014, permission for reuse: number IEA-19–008), SES1 is initially the sum of family possessions, such as books in the child's room, IT equipment, second homes and cars. Home possessions (SES1) were coded as 1 if possessing and 0 if not. SES2 offers choices of seven education levels and are later converted into highest years in education (Xie et al., 2017, p. 93). Parents' levels of education (SES2) were primarily coded sequentially as 1–7 and recoded as the number of years: 1 = primary or below = 6 years, 2 = lower secondary = 9 years, 3 = upper secondary or professional education = 12 years, 4 = associates = 15 years, 5 = bachelors = 16 years, 6 = masters = 19 years, 7 = doctorate = 22 years (Xie et al., 2017, p. 93). SES 3 offers seven career statuses for students to choose (Li et al., 2016, p. 77). Parental career types (SES3) were coded: 1 = no job, 2 = farmers, 3 = non-agricultural workers, 4 = self-employed, 5 = ordinary civil service, 6 = professionals (such as teachers, lawyers, doctors and researchers), 7 = corporate or government leaders (Li et al., 2016, p. 77). Typical examples of career types were provided for children to make choices. On site explanations were offered when the survey papers were delivered in person by the first author in a friendly and neutral tone.

The SES index is generated after a perfect model fit of the SES as a latent variable with the data in the confirmatory factor analysis (CFI = 1.000, TLI = 1.000, RMSEA = 0.000, SRMR = 0.000). The index has a mean of 35.24 and standard deviation of 7.40 on the SES continuum, ranging from 13.62 to 52.67.

3.6.2 Measuring affective learning outcomes in mathematics

Measuring affective learning outcomes allows us to hear the inner voices of the learning and schooling child. More specifically, to understand children's perceptions of maths, teaching and schooling, we utilised three scales from TIMSS 2015 (IEA, 2014) with reuse permission: (1) students' ATM (ASBM01A-I); (2) students' sense of school belonging (SBL) (ASBG11A-G); and (3) students' views on mathematics teaching engagement (EGM) (ASBM02A-J).

3.6.3 Measuring metacognitive learning outcomes in mathematics

For metacognition, we adapted the instrument developed by Sperling et al. (2002), Jr. MAI, by setting it in the mathematical context. Since the target age groups were Grades 5 and 6, the version B was utilised. The original instrument has been field tested by Sperling et al. (2002) and Ning (2016), with the latter conducted in Singapore where 79% of the population are Chinese. In the MasterMT project, the adapted Jr. MAI (v.B) was translated into Chinese and then back to English to achieve the backtranslation validity. Factor analysis and a reliability test (Cronbach's α) were carried out. The factors loaded exactly onto the two intended factors which together explained 45.6% of variance, with α being .907, .841 and .844 for Jr MAI (v.B) as a whole as well as KC and RC as independent constructs respectively.

3.6.4 Measuring cognitive learning outcomes in mathematics

After a pilot with three classes in a participating school, we finalised a set of 28 items for the assessment of student learning outcomes in mathematics. These items are organised in three booklets, G5A, G5B and G6, which share seven common items between them for test equating purposes.

Focusing on rational number sense and proportional reasoning (Figure 3.3), the test items come from the *Decimals*, *Fractions* and *Ratio and Proportion* tests developed by the historical CSMS project in England (Hart, Brown, Kerslake et al., 1985a, 1985b; Hart, Brown, Küchemann et al., 1981). A selection of the CSMS items were also utilised by the Improving Competence and Confidence in Algebra and Multiplicative Structures (ICCAMS) project three decades later (Hodgen et al.,

Decimal	*Item examples*
Have a fundamental understanding of the infinite feature of rational numbers;	D03
Be proficient in calculating or estimating division answers that are not integers but decimals, no matter the dividend is greater (item) or smaller (items and) than the divisor;	D04 D05 D06
Demonstrate the fluency in making conversion between different decimal places, for example, tenths to hundredths (item);	D02
Write a decimal according to the verbal expression of its place value, particularly when it is expressed in such a way that the number of units on the place exceeds 10, for example, writing 11 tenths as 1.1 (item D01).	D01
Fraction	*Item examples*
Apply the knowledge of equivalent fractions (items);	F05–7 F08–9
Model and solve real-world problems using subtraction of fractions with different denominators;	F03
Model and solve real-world problems by dividing fractions that differ in both numerators and denominators;	F10
Model and solve real-world problems by dividing two mixed fractions;	F04
Reason abstractly about fractions in an algebraic situation and provide the necessary condition that satisfies the situation.	F11
Ratio & Proportion	*Item examples*
Identify and apply ratios that are much harder than 2:1 or 1:2 and are set in real-world contexts, for example, where two terms of a ratio are enlarged or shrunk by a fraction (rather than a whole number) with its numerator and denominator being coprime but not 1 or by a fraction that can be simplified as such;	R09 R10
Reason proportionally, use ratio to model real-world problems and ultimately solve the problems;	R02–4 R06 R10
Model and solve real-world problems in more complex situations where three or more quantities are interrelated proportionally.	R02–4 R08

FIGURE 3.3 Rational number sense and proportional reasoning: Assessment framework

Source: Hart et al. (1985a, 1985b); reuse permitted by the ICCAMS project PI, Professor Jeremy Hodgen.

2010). A significant decline in rational numbers, proportional reasoning and algebraic thinking was found amongst a representative sample of English secondary students. Utilising classical items would offer opportunities for us to see student performance through longitudinal/cross-sectional lenses and to understand findings with former studies in mind, even though the focal populations tend to be different in educational research. Conscious of the differences of populations, times and contexts between studies, we see the CSMS results as a benchmark.

In the CSMS project, there were six levels in the decimals (D) test, four levels in two fractions tests – Fractions (F) 1 and 2 – and four levels in the Ratio and Proportion (RP) test (Hart et al., 1985b). In the current study, fifth graders attempted seven decimal items at Levels 5 and 6. Both fifth and sixth graders were given a Level-4 item from F1 (or Level-3 of F2) and one Level-4 item from F2. Besides, fifth graders also were given five other fraction items including four Level-3 and one Level-4 items from F1, whereas sixth graders also did one Level-2 item (as a condition for a Level-4 item) and two Level-4 items from F1 as well as one Level-4 items from F2. All students who took the test in Grades 5 and 6 had tried five items from RP, including one Level-2 item (as part of the item series on 'eel feeding') and four Level-3 items.

Test results were primarily coded with an adapted version of the marking scheme of the CSMS tests (Hart et al., 1985b). The marking scheme provides rich codes for each item on multiple answers (correct or wrong), and possible cognitive processes behind the answers were diagnosed through the CSMS interviews. Thus, there is an explanation for a specific answer to a specific item. A few new codes were added due to their regular occurrences. On each test item, the multiple coding approach generates rich patterns of answers amongst all examinees.

Three undergraduate students enrolled on the primary maths programme at the first author's home institution were recruited to grade and code the maths test, according to the marking scheme (Hart et al., 1985b). The three undergraduates passed the inter-rater tests before the formal coding process. Two half-day training sessions were delivered by the first author, and each was followed by a round of ratings on 10 to 12 test papers. The first round of ratings on ten test papers only saw disagreements between two research assistants on one item per each of three papers; after subsequent discussions and explanation, a perfect inter-rater agreement (pairwise) was reached between the three research assistants on 12 papers (Cohen's kappa: $\bar{\kappa}_1 = .953$, $SD_{\kappa 1} = .08$; $\bar{\kappa}_2 = 1$, $SD_{\kappa 2} = 0$). The multiple codes for each item were recoded dichotomously for the Rasch model, with 1 representing a correct response and 0 otherwise.

In R and RStudio (R Core Team, 2021; RStudio Team, 2021), We utilise the weighted likelihood estimation method in the R package, TAM (Kiefer et al., 2021), to estimate the item and person parameters within each of the three tests (G5a, G5b, and G6) separately. The results indicate an acceptable fit of each test's data to the Rasch model, given that the weighted mean square (infit MNSQ) values (M = 0.99, SD = 0.1) were all within the interval of [0.7, 1.3] (Bond & Fox, 2015).

The person scores initially calibrated within each of the three tests were shifted onto the same scale of G5B via the seven link/anchor items (Wu et al., 2016).

We analyse the patterns of the cognitive learning outcomes by item (facilities), topic (level proficiencies) and test (student performance). An item's facility was the correct rate (%) of the item. For the level proficiency of the three topics – decimals, fractions and ratios and proportions – we refer to the CSMS criteria. In the CSMS study, there were four hierarchical levels for fractions and ratios and six for decimals. According to the marking schemes (Hart et al., 1985b), we calculated the correct rates of individual students on each level and recoded them dichotomously as 1 if the correct rate was equal to or higher than the particular level criterion and 0 if not.

To test the effects of age, gender and SES on the maths test, we run primary analyses of the data using pairwise correlations and *t*-tests (gender). These shed light on the selection of control variables in our subsequent analyses using multi-level and structural equation modelling.

3.7 Modelling the teaching-learning mechanisms

The ideal way of validating educational effects is randomised controlled trials (RCTs), and correlational studies cannot establish causal relationship between variables. Nevertheless, in studies where RCTs are hard to undertake, the next option is to apply more complexed statistical methods, such as single- or multilevel structural equation models (SEMs/MSEMs), to reflect the complexed reality.

In the MasterMT project, multilevel and structural equation models were run to examine the relation between teaching and learning. For the MLM, two-level models were run to estimate the effects of teacher-level predictors on three types of learning outcomes – affective, metacognitive and cognitive outcomes. The SEMs/MSEMs are run to measure direct and indirect effects of teaching on the three types of learning outcomes.

All models were run in R and RStudio (R Core Team, 2021; RStudio Team, 2021). The two-level linear mixed models are estimated with the *lme4* package in R, using the maximum likelihood method for model fit and Satterthwaite's method for *t*-tests (Bates et al., 2015). For each scale/latent variable, we check the internal reliability first and then run a CFA with the lavaan package (Rosseel, 2012). The package was also used to test the hypothesised mediating models (SEMs or MSEMs).

The fit of CFA and SEMs was checked with the following stand-alone fit indices: the Chi-square test, the comparative fit index (CFI), the Tucker-Lewis index (TLI), the root mean square error of approximation (RMSEA) and the standardised root mean square residual (SRMR). The Chi-square test should be not significant but it can be sensitive to a large sample size. The CFI would be considered good if ≥ 0.95 or acceptable if ≥ 0.90 (Hu & Bentler, 1999; Kline, 2005). The TLI would be considered good if ≥ 0.90 (Bentler & Bonett, 1980). The RMSEA would be considered good if ≤ 0.06, acceptable if ≤ 0.08, and marginal if ≤ 0.10 (Hu & Bentler, 1999; McDonald & Ho, 2002); RMSEA is likely to be greater than 0.10 if the *df* is small (Kenny et al., 2015). A SRMR would be considered good if ≤ 0.08 (Hu & Bentler, 1999). We chose to consult indices other than the Chi-square test or RMSEA in the following two situations: (1) a large sample size (e.g., measurement at the student level) with an insignificant Chi-square test; (2) a small *df* with a large RMSEA.

3.8 Studying the development of master mathematics teachers

Beyond the anticipated excellence, we are interested in the development of master mathematics teachers – something might be better interpreted through the qualitative ethnographic approach. More specifically, we want to know the trajectories and current status of master teachers' PD. To capture their PD trajectories and PD in action, we collected extra data in addition to the classroom data. In the second semester following our major data collection in classrooms, we asked teachers to reflect upon their journey towards master teachers and the development of their master teacher studios should they be running one and have time to write about it. We tapped into the PD reality by participating in two TRG meetings in Ms Q's school (Chapter 2) and teaching research conferences at the municipal, provincial and national levels.

The two parts of data allow us to see master teachers' PD in a broader sense from two major angles: (1) from the teachers' points of view, with teachers telling their stories; (2) from the researchers' points of view, with us seeing PD in action. As part of the PD trajectories, seven teachers also share with us the development of their MTSs and the activities hosted or attended by the MTSs. These complement both the PD-trajectories and PD-in-action data, allowing us to have deeper understanding of master teachers' development and their leading role in fellow teachers' PD from the bottom up.

The teacher PD trajectories data were collected in April 2020 through a semi-structured open-ended survey with teachers. In the Word-based survey, we asked teachers to: (1) write a short bio about themselves; (2) reflect on their PD trajectories through which they developed into a teaching master; (3) list teaching research events that they took part in from January 2018 up to the time of the PD survey. An optional survey was also designed in a similar manner aiming for those who were hosts of MTSs at the time. The MTS survey was also of three sections: (1) introduction of the MTS; (2) the development of the MTS and major events/milestones in its development; (3) events/activities of the MTS over the same time span. Two survey files were sent out to all 70 teacher participants. In the middle of the pandemic, we appreciated that 38 managed to respond to the PD trajectories section, with seven of them having also completed the MTSs section.

Before the formal content analyses of the PD trajectories data, a series of procedures were conducted to make the data unified across cases. The raw data were in Chinese and re-formatted by: (1) keeping all the font to be 11-point SimSun (简宋); (2) keeping the line space to be 1; (3) having the headers (as put in the blank copy) removed; (4) replacing the three headings (survey prompts) with two horizontal lines (set as double space) to keep the documents in three parts that were solely materials written by each teacher; (5) setting the margins as normal (2.54 cm on each side).

To get an overview of most frequently mentioned words by the teachers, our analyses of the written accounts of PD trajectories start with a word frequency or word cloud analysis using the software ATLAS.ti. To get the word cloud, we set the frequency threshold at 100 and at a maximum number that would give us three most frequently used words/phrases. This overview in fact is captured at the word level so we consider this a *micro*analysis of the data.

Then, our major attention is given to in-depth analyses of themes and patterns embedded in the data through the grounded theory approach (Glaser & Strauss, 1967) which allows the themes, patterns or theory to emerge from the data (Cohen et al., 2018). Using the constant comparison approach, we are able to combine the strengths of two approaches to qualitative analyses: the explicit coding and theory development. In seeking the overarching meanings/themes, we are able to complete a *macro*analysis of the data.

3.9 Multiple perspectives, multiple layers

At the macro level of the project, multiple perspectives interweave in multiple layers. Unstructured lesson observations generate researchers' perspectives; teachers give their views about their own teaching; about one of the case study teacher's lessons, an array of diverse perspectives was uttered by teachers' colleagues in the same department and the municipal teaching research official. These are packed with multiple perspectives about masterly mathematics teaching, collected from professional learning communities specialised in primary mathematics. Moreover, one of the mathematics survey sections also captures students' perspectives regarding schooling, teaching and learning.

Immediately after the lesson, the teacher completed an interview on the lesson. No set questions were given to teachers in the audio-record post-lesson interview, and teachers were free to talk about anything they would like to say about the lessons observed and mathematics teaching and learning in general. The intention was to generate key aspects of maths teaching that teachers were most concerned with when reflecting upon the lesson they just delivered and talking about mathematics teaching and learning in general. This, together with the open-ended survey with teachers regarding their PD trajectories, lets the master minds speak for themselves.

In addition, teachers also completed a questionnaire asking about their background information, teacher knowledge structure (TT2G12), teaching beliefs

(TT2M14) and self-efficacy (TT2M15), with the latter three aspects captured by three of the TALIS 2013 scales (OECD, 2013). The integration of teacher questionnaire and interview data yields insights into teachers' teaching beliefs, SMK, PCK and metacognition (MC). These are further connected with (1) initial findings of both structured and unstructured observations of lessons and (2) relations between teaching and learning. Through participatory observation of the case study teacher and the PD events at various levels, we as researchers also gain ethnographic insights into 'the culture, values, beliefs and practices' (Cohen et al., 2018, p. 292) of master mathematics teachers and PD in action of mathematics teachers in China.

3.10   Multiple evidence, one reality: ready for the MasterMT journey

Ultimately, at the macro level of the project, multiple measurements at multiple levels join multiple perspectives through multiple layers to form the one holistic reality where optimal teaching and learning is expected to happen.

About mathematics teaching, we have: (1) structured observation using three instruments, the OTL, the ISTOF and the MQI; (2) student-perceived teaching engagement collected as part of the student questionnaire; (3) teacher self-comments on the observed lesson and maths teaching in general during the post-lesson interview; (4) teachers' teaching beliefs collected with the teacher questionnaire; (5) teacher's and colleagues' comments on the lessons delivered during the teaching research meetings/conferences; (6) in-depth case study of a master teacher's teaching over 40 school days; (7) our unstructured observation of the lessons as researchers.

About learning, we have: (1) affective learning outcomes as measured with the TIMSS 2015 items; (2) metacognitive learning outcomes as measured with the Jr MAI questionnaire; (3) cognitive learning outcomes in mathematics as measured with the test; (4) multicategories of learning-in-action captured by our lesson observations.

About the teaching-learning mechanisms, we have: (1) multilevel modelling of teaching effects on three types of learning outcomes; (2) multilevel structural equation modelling of the direct and indirect effects of teaching on learning; (3) teacher beliefs on what works and how teaching works on learning (interview); (4) the observed interaction between teaching and learning during the teaching research sessions and teachers' individual and collective interpretation of teaching and learning just observed; (5) our unstructured observation of interaction between teaching and learning.

The three major parts of the analyses were initially reported separately in accordance with the convention of each specific research method and then systematically integrated once the preliminary findings emerged, such that we could gain insights into connections between and beneath the findings.

The case study and PD data provide rich and vivid answers to the fundamental question we inevitably want to ask: what makes master teachers and masterly teaching? The longitudinal ethnographic study of Ms Q's work generates insights into a master teacher's everyday teaching and work, her beliefs and techniques in improving learning at the 'nano-level', her role in leading teachers' PD and so on. The analyses of PD-trajectories and PD-in-action data allow us to see the development of master mathematics teachers over time and in real time from a wide angle.

On an abstract sense, the methodology here addresses two fundamental questions in education research: What exactly is the essence of reality to be studied? How can the reality be studied? Our answer to the first question is that the reality is neither objective nor subjective – it is both in one, which in itself gives an overarching answer to the second question. It is only through thorough integration that the reality can be known with immense authenticity. Taken together, the methodology is helpful in generating thorough and robust evidence for research, practice and policy-making in mathematics education, as it makes the best use of both quantitative and qualitative methods and data. Having laid out the paradigmatic ground and methodological plans for the project, we are ready to analyse data and present our major findings in Chapters 4 through 9, in addition to the case study in Chapter 2.

4

MASTERLY MATHEMATICS TEACHING IN EVERYDAY CLASSROOMS

Chapter overview

- Masterly teaching through the three quantitative lenses
- The quantity of teaching measured with the OTL
- The quality of teaching measured with the ISTOF
- The quality of maths teaching measured with the MQI
- The ISTOF and the MQI as latent teaching variables
- Masterly teaching through the qualitative lens
- Model the way for a shared discourse
- Multiple representations, one mathematical fact
- Not moving on until the class have reasoned in-depth the very essence of the task/topic
- Variation as scaffolds for fundamental understanding
- Lessons built on a variety of student contribution in an optimal sequence
- Constant abstracting and generalising
- Key structure of mathematics as the core of each lesson
- Teacher gradually unfolding the essence of knowledge on the board
- Eight other features of the masterly teaching

> Everybody told me before I ever studied teaching in Japan that teaching in Japan was rote. So, I went and studied elementary schools in Japan, and it doesn't look rote. Whatever it is, it's not rote.
>
> – James Stigler

DOI: 10.4324/9781003127925-4

This quote is from a talk that James Stigler – the author of *The Learning Gap* (Stevenson & Stigler, 1992) and *The Teaching Gap* (Stigler & Hiebert, 1999) – once gave about mathematics teaching in Japan.

We share Stigler's view when talking about mathematics teaching in China. It echoes the findings of the international study on the effectiveness of mathematics teaching in average urban classrooms (Miao & Reynolds, 2018) and the findings about the teaching practice of Chinese master mathematics teachers in the current study.

As discussed in Chapter 3, lessons are video-recorded and analysed both quantitatively and qualitatively. Quantitative analyses are conducted with three observation systems: the OTL adapted from Reynolds et al. (2002), the ISTOF developed by Teddlie et al. (2006) and the MQI developed by Hill and Ball (2004) and colleagues on the Learning Mathematics for Teaching (2011) project. The analyses are carried out on the condition of acceptable inter-rater agreement within the project team (see Chapter 3). Qualitative analyses are conducted through the grounded theory approach. Ultimately, we hope to study the lessons to their essence by focusing on both visible and invisible aspects of teaching. Integrating quantitative and qualitative approaches, we are able to do so.

4.1 Masterly teaching through the three quantitative lenses

Across the 70 classes, the average duration of observed lessons is 43 (*SD* = 3) minutes, with the class size averaged to 45 (*SD* = 9). As follows, we will present detailed results of the quantitative measurements of teaching, using the three observation systems: the OTL, the ISTOF and the MQI. The OTL measures the quantity of teaching in six aspects: the proportion of lesson time spent on five types of classroom activities and student time on task. The overall quality of teaching and the overall quality of mathematics teaching are measured on a scale of five with the ISTOF and the MQI respectively.

4.2 The quantity of teaching measured with the OTL

The OTL measures the proportion of lesson time spent on whole-class interaction (OTL1), whole-class lecture (OTL2), individual/group work (OTL3), classroom management (OTL4) and partial class interaction (OTL5), with an extra focus on the average proportion of student time on task in the lesson (OTL6).

OTL1 vs. OTL2 vs. OTL5. The major part of each lesson is given to *whole-class interaction* (80.7 ± 8.1%), ranging from 62.2% to 100.0%. More than half of the teachers (62 of 70) allocated 70% or more of lesson time for interaction with the whole class. *Whole class lecturing* was observed in only one of all the classes in Anhui, on one occasion lasting for 0.6% (14 sec) of the lesson time (42 min 21 sec). No time across all the lessons was spent on class management. *Partial class interaction* is not observed in any of the lessons, as found previously in maths lessons in Nanjing (Miao & Reynolds, 2018). There is child-centred exploration into mathematics at the whole-class level both independently and collectively rather than at the individual level.

OTL3. Across the classrooms, the proportion of lesson time on *independent seatwork* ranged from null to 33.4%. In slightly over 8/10 of the lessons (57 of 70), teachers allocated less than 20% of lesson time for children to work independently in an immediate manner – usually within a minute. In 18 of 70 lessons, the coverage of *groupwork* added up to more than 10% of the lesson time, with the maximum proportion being 19.9%. In total, the three subcategories of OTL3 (student work independently or in groups of two or four) occupied 21.3% to 37.8% of lesson time in 28 out of 70 classes, 10.4% to 19.9% of lesson time in 36 classes, and 5.7% to 9.8% of lesson time in five classes, with one Grade 2 class spending all time on whole-class interaction and no time on any other OTL activities. In 32 of 70 classes, teachers organise *paired groupwork* with the total proportion of lesson time ranging from 0.7% to 16.7%. In 27 lessons, teachers organise *collaborative work in groups of 4*, with 12 of them spending a total of 10.5% to 19.9% of the lesson time on it and others 1.4% to 8.7%. Six lessons consisted of both forms of groupwork. Although the number of classes having paired groupwork is more than the number of those having students working in groups of four, the proportion of lesson time seems to suggest that teachers who arrange groupwork intend to allocate more time for collaboration in groups of four rather than two. Whatever the form is, all OTL3 activities are clearly modelled beforehand, with detailed rules and goals clarified, such that time is spent efficiently and effectively by the children.

OTL4. Almost all of the teachers (66 of 70) do not spend any time in *managing the class*. In four classes, teachers use 0.3%, 0.6%, 1.0% and 5.3% of lesson time to manage the class respectively. The management is very brief and efficient in most cases.

OTL6. Across classes, students are highly engaged, with an average of 99.1% ($SD = 1.4\%$) on task. All students in 36 of the 70 classes were on task throughout the lessons. In 33 classes, the average proportion of students on task was 95.1% to 99.7%. In one class, the proportion was 91.3%.

With the quantity of teaching in mind, we now move on to look at the quality of teaching in general, measured with the ISTOF.

4.3   The quality of teaching measured with the ISTOF

The adapted version of the ISTOF consists of six components. Following the original ISTOF component numbering, they are Assessment and Evaluation (ISTOF1), Clarity of Instruction (ISTOF3), Instructional Skills (ISTOF4), Promoting Active Learning and Developing Metacognitive Skills (ISTOF5), Classroom Climate (ISTOF6), and Classroom Management (ISTOF7). In this section, we detail the results of the ISTOF measurement. Teachers had a mean score higher than 4.1 in each of the six ISTOF dimensions.

Assessment and evaluation (ISTOF1). This dimension sees the highest rating of teaching and smallest standard deviation (*Mean* = 4.60, *SD* = 0.4) in comparison with other dimensions. In assessing children's prior learning and readiness for the new content, teachers demonstrate sound questioning and feedback skills and arrange well-prepared tasks. This dimension is significantly related to children's time on task ($r = 0.34$, $p < 0.01$), perhaps due to the close attention of the teacher to children's state of knowledge and mind that's captured by this dimension.

Clarity of instruction (ISTOF3). Another universal impression across the lessons is clarity (*Mean* = 4.39, *SD* = 0.59). At the macro level, there is a clear sectioning of beginning, main part and closing parts in each lesson. At the meso level, each part of a lesson is clearly structured with carefully sequenced tasks and activities. At the micro level, the chains of questioning, response and feedback lead each utterance of the teacher or student(s) to the next utterance, which guides the logical flow of the class discourse towards the development of mathematical cognition and metacognition. A few questions from the teacher serve to quickly diagnose the starting point of students' learning journey. Teachers are good at unfolding the lesson title by initiating a short dialogue on a seemingly irrelevant event from the real world or from student prior knowledge. Scaffolding takes a variety of forms to mimic the necessary stepping stones for learners as vividly as possible. Teachers appear to have prepared *key questions* to catalyse desired learning processes. Since

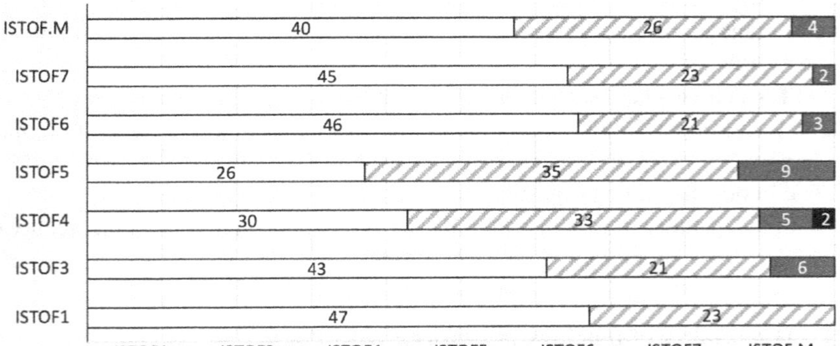

	ISTOF1	ISTOF3	ISTOF4	ISTOF5	ISTOF6	ISTOF7	ISTOF.M
□ Score 5	47	43	30	26	46	45	40
▨ Score 4	23	21	33	35	21	23	26
▪ Score 3	0	6	5	9	3	2	4
▪ Score 2	0	0	2	0	0	0	0

FIGURE 4.1 Distribution of lesson scores on the six 5-point scales of ISTOF

the teaching is clear, children can make sense of the content, hence unlikely to drift off. Unsurprisingly, the clarity of instruction is significantly correlated with student time on task ($r = 0.38, p < 0.01$).

Instructional skills (ISTOF4). About 9/10 ($n = 63$) of the teachers reached a score of 4 or higher on instructional skills. Their lessons are highly interactive, posing questions and tasks arousing active response and participation from the class. Teaching in their classes takes various forms, so that children's attention is constantly drawn towards the flow of the mathematical content along the timeline. The ISTOF4 score is positively correlated to student time on task ($r = 0.24, p < 0.05$).

Promoting active learning and developing metacognitive skills (ISTOF5). This is the most challenging dimension in ISTOF measured with ten items and four indicators. Almost 9/10 ($n = 61$) of the teachers reached a raw score of 4 or higher on a scale of 5 in this dimension, with nine lessons having a score of 3. This dimension is also a positive correlate of time on task ($r = 0.27, p < 0.05$).

Classroom climate (ISTOF6). Classroom climate is one of the three dimensions with raw scores averaged to about 4.5 and a small *SD* (0.5). It is universal that all lessons observed have a positive climate for quality teaching and learning to happen. Almost all lessons ($n = 67$) are rated 4 or 5 on a scale of 5 in this respect. The classroom climate is not correlated with time on task ($r = 0.16, p = 0.18$).

Classroom management (ISTOF7). This dimension is about quality of class management (*Mean* = 4.57, *SD* = 0.46). Teachers seem to have mastered the managing skills without observable managing behaviours. Almost all ($n = 68$) of the 70 lessons have reached 4 or 5 on the scale of 5, with two lessons being rated 3 in this dimension. Students become part of it. In quite a lot of classes, students are able to call upon their peers, comment on their presentations or answer the questions or problems they have just posed. Such student-student interaction at the whole class level seems to be a convention long formed in the class. Their good behaviour could not be just out of natural instinct but is more likely due to long-term effort into the collective training of such discipline and determination in keeping everything relevant to mathematics and everything in order. It is thus not just the teacher who manages the class well. It is both the teacher and students who work towards a shared vision. Of course, the teacher must be the primary initiator. Unsurprisingly, the quality of classroom management is positively correlated with student time on task ($r = 0.43, p < 0.01$).

4.4 The quality of mathematic teaching measured with the MQI

In this section, we detail the quality of mathematics teaching measured with the ten MQI whole-lesson items (Hill & Ball, 2004; Learning Mathematics for Teaching, 2011). More specifically, we look at the *lesson efficiency* (MQ1), *denseness of mathematics* (MQ2), *student engagement* (MQ3), *richness of mathematics* (MQ4), *remediation of student difficulty* (MQ5), *using student ideas* (MQ6), *clarity and accuracy of mathematics* (MQ7), *quality tasks and activities, CCSSM* (Common Core State Standards: Mathematics) *aligned contents* (MQ9), and *overall mathematical quality of the instruction* (MQ10).

Teachers have an average raw score of 4 or higher in all MQI items except the MQ9 measuring the extent to which the lessons generate CCSSM practices. Over two-thirds of the teachers make the desired CCSSM practices happen in their lessons. On all MQI whole-lesson items except MQ4 and MQ9, over 4/5 of teachers scored 4 or 5 out of 5. On the denseness of mathematics (MQ2), 58 teachers got a full score and nine got a 4, which places all 67 teachers at 4 or 5 on a scale of 5.

Lesson efficiency (MQ1). A common feature across the classrooms is lesson time being used properly and efficiently (MQ1). The 70 lessons have a mean score of 4.66 ($SD = 0.61$). In 48 lessons, the time efficiency is rated the highest (i.e., 5); 65 of the lessons got a score of 4 or 5 in this regard, with only five lessons being rated as 3.

Denseness of mathematics (MQ2). All classes delve themselves deep into mathematics. Because of the rather high mathematical density, the 70 lessons have a mean score of 4.87 ($SD = 0.34$) on MQ2. This is the only item that saw all teachers get a score of 4 or 5. Of the 70 lessons, 58 got a score of 5, with nine being rated as 4. Across classrooms, students and teachers are deeply immersed in the subject. No time is wasted on transitions between activities or issues irrelevant to mathematics. Teachers are conscious of the flow of content in the class and in children's cognition.

Engagement of students (MQ3). The extent to which students are engaged in the lesson is consistent with the OTL6 measure, time on task. Over half of the 70 classes have a full score on engagement of students, with no off-task phenomena observed. In fact, engagement tends to be well maintained in the majority of lessons even in average classrooms in urban China (Miao & Reynolds, 2018). In almost 9/10 of the lessons, engagement is rated as 4 or 5. The average score is 4.43 ($SD = 0.73$) for the 70 lessons.

Richness of mathematics (MQ4). In 7/10 of the classes, there is a consistent existence of rich mathematics (*Mean* = 4.04, *SD* = 0.87). The MQ4 sees 26 lessons being rated as 5 and 23 as 4. Across the lessons, there is a strong emphasis of building connections between mathematical concepts, between representations, between multiple solutions, between mathematics and life. Students actively join the mathematical discourse to explain their ideas and comment on other students' ideas. Teachers gently nudge the learning minds to think about the *how* and *why* questions and make sense of the seemingly solid facts/solutions. Students' reasoning is articulated with rigid mathematical language.

Getting to the bottom of learning difficulty (MQ5). In over 8/10 ($n = 57$) of the 70 classes, teachers have scored 4 or 5 for addressing learner difficulty; in 13 classes, the teaching is rated as 3. The mean score of MQ5 is 4.14 ($SD = 0.71$) for the 70 lessons. It is clear that the centre of each teacher's attention is on the status and development of knowledge in children's mind. They all care greatly about what can promote or hinder learning. Given the fact that a big chunk of lesson time is

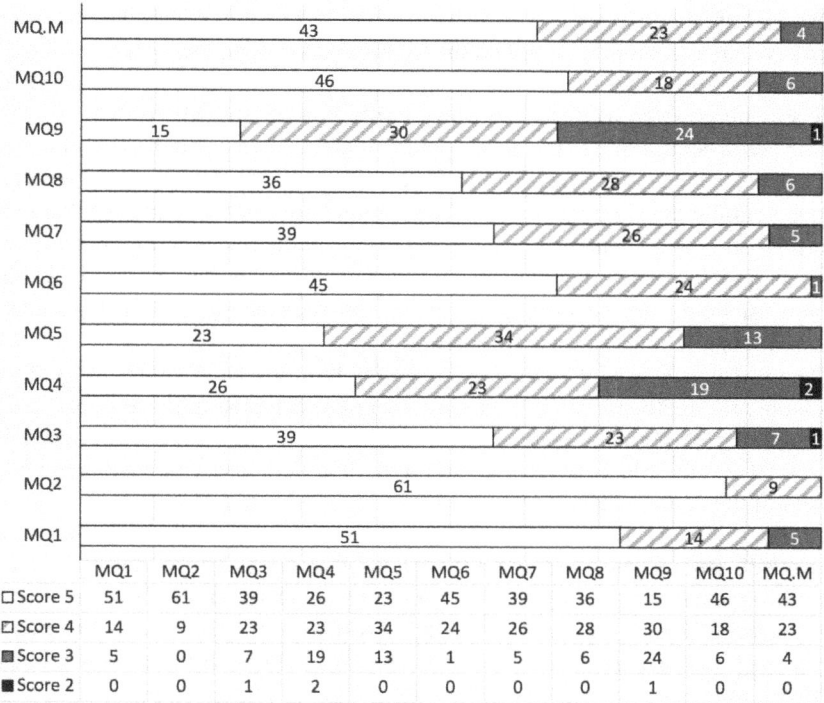

FIGURE 4.2   Distribution of teachers scoring on the 5-point whole-lesson scale of MQI

Note: MQ1 to MQ10 refer to the MQI whole-lesson items 1 to 10; MQ.M = mean score of the MQI whole-lesson measure. MQ1 = time efficiency. MQ2 = denseness. MQ3 = engagement. MQ4 = richness. MQ5 = addressing learner difficulty. MQ6 = using student ideas. MQ7 = clarity and accuracy. MQ8 = quality tasks and activities. MQ9 = CCSSM aligned contents. MQ10 = overall mathematical quality of the instruction.

given to whole-class interaction, teachers tend to address learner difficulty during this type of lesson time and had often already made quite accurate predictions of typical obstacles. Hence, at a deeper level, you could find, similar to managing the class without apparent managing behaviours, quite often, teachers address learning difficulties in an appropriate manner without explicitly pointing them out. Teachers tend to naturally weave the elements of major teaching purposes into various interaction and activities. The more apparent way of addressing difficulty is to present certain misconceptions embedded in children's work/ideas to the whole class and resolve them through peer reviews and collective summaries. Commonly seen forms of doing so include: (1) typical work samples presented through overhead projector, and (2) having student work/solution written on the blackboard.

Using student ideas (MQ6). In 69 of 70 classes, teaching and learning involves a great deal of student response and their work as teaching and learning materials, which gives them a score of 4 or 5 on the MQ6. On the MQ6, the 70 lessons have a mean score of 4.63 ($SD = 0.52$). Teaching in 42 of these classes is rated as 5, with 24 classes being rated as 4. In fact, the contribution of students to the class discourse plays a central role in the development of key knowledge and skills amongst themselves. Using student ideas is also a way of getting students engaged with the content, echoing the significant strong correlation between the ratings of this item and MQ3 on engagement ($r = 0.70$, $p < 0.01$). Frequently inviting contribution from the students also extends the scope of collective thinking and enriches the lesson substantially, hence a strong correlation between the ratings of using student ideas and lesson richness ($r = 0.58$, $p < 0.01$).

Clarity and accuracy of mathematics (MQ7). In more than 9/10 ($n = 65$) of the 70 lessons, the mathematics is clear and accurate, demonstrating teachers' deep understanding of both mathematics and students. The 70 lessons average 4.49 ($SD = 0.63$) on the MQ7. In addition to clarity, the item also emphasises the correctness of the content delivered.

Quality tasks and activities (MQ8). Similarly, in more than 9/10 (64/70) of the lessons, well-tailored tasks and smoothly organised activities lead to thorough development of mathematics amongst the students. On the MQ8, 64 lessons score 4 or 5, and six score 3. The average score of MQ8 is 4.43 ($SD = 0.65$) across the 70 lessons.

Common core aligned content (MQ9). About two-thirds of the classes are featured with those practices demanded by the CCSSM. With a mean score of 3.84 ($SD = 0.77$) for 70 lessons, this is the item that appears to be least realised amongst teachers, perhaps because various types of practices are expected by the CCSSM. All teachers do better or brilliantly on one or some of the CCSSM practices but 15 of the 70 teachers perform the best on all eight practices, with a full score of 5; 30 teachers do relatively well, with a score of 4. In all classrooms, students are called upon to explain their thinking or solutions about the mathematical content under class discussion, to comment on their peers' work or ideas. Students have the opportunity to reason mathematically in the whole class or with their peers in groups, with the teachers taking the position of asking questions. In some of the classes, children are encouraged to pose problems or questions for their peers to tackle or respond to. In most classrooms, students have the opportunity to communicate with their peers in groups or in the whole class about the subject matter. In most classrooms, the tasks given expect students to give solution procedures with a thorough understanding of the underlying concepts. In almost all classrooms, the content is set in contexts; those few not contextualised are due to the nature of the mathematical content being abstract in itself, as it would be in mathematics. Many lessons follow the loop of *contextualised problems – abstracting the mathematics behind the problem – applying the mathematics learnt to problems set in the real world.* Most classes have a strong feature of sense making in mathematical tasks amongst students. The flow of class discourse is moving forward as the whole class strive to tackle the problems at hand. The determination is that, with flexible and thorough thinking, the solution must be found.

Overall MQI (MQ10). More than 9/10 ($n = 64$) of the 70 lessons are rated as 4 or 5 in their overall quality of mathematics in teaching. On the MQ10, all lessons average 4.57 ($SD = 0.65$). Teachers demonstrate not only a profound understanding of both mathematics and students but also a strong competence to build connections between the two. Mathematical knowledge and skills are developed steadily, with rich materials, diverse activities, deep thinking and reasoning, and persistent arguments spreading optimally across the lesson time.

In terms of teaching quality, not all teachers scored the highest in all dimensions of ISTOF and MQI, but some of them did demonstrate top effectiveness and quality of maths teaching in all aspects. Amongst the master mathematics teachers, though there is still variation of lesson time allocations, the MasterMT study replicated the positive correlation found previously between the quality of teaching and student time on task (Miao & Reynolds, 2015, 2018; Muijs & Reynolds, 2000). Classes with a higher percentage of time on task tend to score higher on the MQI ($r = 0.34$, $p < 0.01$) and four of the six ISTOF dimensions: (ISTOF1) *assessment and evaluation* ($r = 0.34$, $p < 0.05$), (ISTOF3) *clarity of instruction* ($r = 0.38$, $p < 0.01$), (ISTOF5) *promoting active learning and metacognitive skills* ($r = 0.27$, $p < 0.05$) and (ISTOF7) *class management quality* ($r = 0.43$, $p < 0.01$). High quality of teaching keeps children engaged in the thinking and learning of the content. The MasterMT has further proved the positive correlation between time on task and the quality of either subject-general or subject-specific aspects of teaching. It is natural to see that classroom climate is found to be positively correlated with the mathematical quality of instruction ($r = 0.85$, $p < 0.01$). In classes with a better climate, students are more engaged with the mathematical content ($r = 0.77$, $p < 0.01$). A high quality of class management leads to a high level of engagement with the mathematical content ($r = 0.76$, $p < 0.01$), in addition to the high proportion of children time on task ($r = 0.43$, $p < 0.01$). A class scored higher in management tends to score higher on clarity ($r = 0.801$, $p < 0.01$) and instructional skills ($r = 0.803$, $p < 0.01$) as well. In a lesson like this, the teacher is more likely to promote active learning and attend to students' metacognitive development ($r = 0.78$, $p < 0.01$).

Next, we will explore the latent structure of these two teaching-quality measures, checking the appropriateness of using them as the measurement parts of the structural equation models hypothesised in Chapter 7.

4.5   The ISTOF and the MQI as latent teaching variables

The latent constructs are tested using the CFA. As aforementioned, due to timing issues, three of the 70 classes did not do the maths survey. Considering the SEM analyses would need learning outcomes data collected with the survey, we only use 67 teachers' teaching data for the CFA and SEM analyses.

It is natural to think all ISTOF components and the MQI are each, of themselves, a latent variable (H_{ISTOF1}, H_{ISTOF3}, H_{ISTOF4}, H_{ISTOF5}, H_{ISTOF6}, H_{ISTOF7} and H_{MQI}). It is also possible that ISTOF may converge into a second-order latent variable with items/indictors explaining components and components explaining the overarching ISTOF concept (H_{ISTOF}). Our impression with the ratings is that all classrooms are similar in terms of assessment (ISTOF1), climate (ISTOF6) and discipline (ISTOF7), maths density (MQI2), student engagement (MQI3), maths clarity (MQI7), and CCSSM alignment (MQI9), with most classes scoring quite high on all of these except MQI9 where more variation is observed than other MQI predictors. For this reason, we propose that confirmatory factor analyses may accept the ISTOF and MQI each as a first- or second-order latent variable which consist of items/components apart from those mentioned above, rather than all that were used in the study.

Furthermore, in Table 4.1, with Cohen's r ranging from 0.63 to 0.90, strong pairwise correlations are observed between the MQI and six ISTOF components. This begs the question as to whether both observation systems are affiliated to one higher-order latent variable that reflects teaching quality on both subject-general and subject-specific indicators (H_{IFMQ}).

We test these hypotheses through a series of confirmatory factor analyses, first with the ISTOF1, ISTOF3, ISTOF4, ISTOF5, ISTOF6, ISTOF7 and the MQI each as a first-order latent variable (Table 4.2) and then with the ISTOF (i.e., H_{ISTOF}) and the two systems combined (i.e., H_{IFMQ}) each as a second-order latent variable (Table 4.3).

TABLE 4.1 Zero-order correlations between three observation systems

	M	SD	MQI	IF1	IF3	IF4	IF5	IF6	IF7	OTL1	OTL3
MQI	4.42	0.51	1								
IF1	4.60	0.40	**0.78**	1							
IF3	4.40	0.59	**0.90**	**0.81**	1						
IF4	4.21	0.70	**0.87**	**0.71**	**0.85**	1					
IF5	4.18	0.57	**0.82**	**0.63**	**0.81**	**0.79**	1				
IF6	4.53	0.50	**0.85**	**0.78**	**0.84**	**0.83**	**0.72**	1			
IF7	4.59	0.46	**0.83**	**0.70**	**0.81**	**0.79**	**0.77**	**0.79**	1		
OTL1	0.81	0.08	0.10	0.09	0.03	0.10	-0.05	0.08	0.02	1	
OTL3	0.19	0.08	-0.08	-0.08	-0.02	-0.10	0.06	-0.07	-0.01	**-0.10**	1
OTL6	0.99	0.01	**0.34**	**0.34**	**0.38**	0.22	***0.27***	0.16	**0.43**	0.01	0.01

Note: Coefficients in bold or bold and italics are significant at the level of 0.01 or 0.05 (two-tailed) respectively. IF1 (ISTOF1) = Assessment; IF3 (ISTOF3) = Teaching Clarity; IF4 (ISTOF4) = Instructive Skills; IF5 (ISTOF5) = Promoting Active Learning and Metacognitive Development; IF6 (ISTOF6) = Classroom Climate; IF7 (ISTOF7) = Class Management; OTL1 = Whole-class Interaction; OTL3 = Student Independent or Group Work; OTL6 = Student Time on Task. MQI and ISTOF scales are of five points; OTL measures are percentages ranging from 0 to 1 (i.e., 0% to 100%).

TABLE 4.2 Fit indices for the CFA models of ISTOF components and the MQI

LV model	N	χ^2	df	χ^2/df	p	CFI	TLI	RMSEA	SRMR
MQI_{6i}	67	10.89	9	1.21	0.284	0.992	0.986	0.056	0.044
IF1	67	1.04	2	0.52	0.595	1.000	1.061	0.000	0.027
IF3	67	12.35	9	1.37	0.194	0.987	0.979	0.075	0.036
IF4	67	7.76	5	1.55	0.170	0.989	0.978	0.023	0.023
IF5	67	1.56	2	0.78	0.459	1.000	1.012	0.000	0.022
IF6	67	0	0	NA	NA	1.000	1.000	0.000	0.000
IF7	67	0	0	NA	NA	1.000	1.000	0.000	0.000

Note. LV = latent variables. CFI = comparative fit index. TLI = Tucker-Lewis index. RMSEA = root mean square of approximation. SRMR = standardised root mean square residual. MQI_{6i} = the 6-item model consisting of whole-lesson items 1, 4, 5, 6, 8, and 10.

TABLE 4.3 Fit indices for ISTOF (2nd order) and MQI whole-lesson codes

Model	χ^2	df	p	χ^2/df	CFI	TLI	RMSEA	SRMR	AIC	BIC
MT1	443.919	269	0.000	1.650	0.882	0.868	0.099	0.060	1934.939	2058.401
MT2	327.361	204	0.000	1.605	0.900	0.887	0.095	0.060	1819.038	1927.068
MT3	**204.352**	**148**	**0.001**	**1.381**	**0.942**	**0.934**	**0.075**	**0.058**	**1613.893**	**1706.490**
MT4	**10.885**	**9**	**0.284**	**1.209**	**0.992**	**0.986**	**0.056**	**0.044**	**597.634**	**624.091**
MT5	444.95	270	0.000	1.648	0.878	0.864	0.098	0.065	2103.027	2224.285
MT6	760.85	427	0.000	1.782	0.831	0.816	0.108	0.065	2418.652	2570.776

Note: MT1 = 2nd-order CFA on IF134567. MT2 = 2nd-order CFA_IF13456. MT3 = 2nd-order CFA_IF1345. MT4 = CFA_MQ-1456810. MT5 = 2nd-order CFA_IF1345MQ. MT6 = 2nd-order CFA_IF134567MQ.

ISTOF6 and ISTOF7 converged into saturated models, each indicating a perfect model fit (Table 4.2). Based on both our impression of the observation and the CFA results of ISTOF6 and ISTOF7, we proceeded to run the proposed second-order CFA for ISTOF, including and excluding the two components for model selection.

As we anticipated, the MQI has a better fit with six of the ten items (#1, 4, 5, 6, 8 and 10). The CFA results suggest a good model fit (Table 4.2 and Figure 4.3) of the six-item MQI construct.

The CFA models on various combinations of the constructs show a good fit for each of the two proposed combinations of ISTOF and MQI models to the data. The first is the second-order CFA model on ISTOF with four components – ISTOF1, ISTOF3, ISTOF4 and ISTOF5. The second is the first-order CFA model of the MQI consisting of six items (#1, 4, 5, 6, 8, and 10). The four items that were excluded from the latent variable construct were, as discussed in the descriptive statistics section, quite similar across classrooms, which might be the reason for

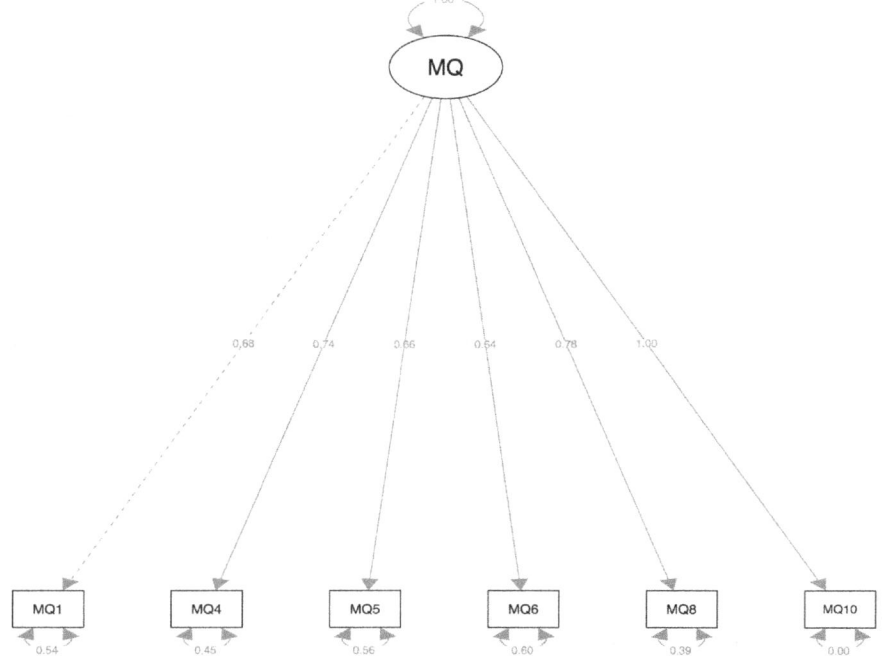

FIGURE 4.3 A CFA model of the MQI with six items

not contributing to the latent construct well. However, we chose to parcel all ten MQI items and use the mean as a MQI score of each teacher in the multilevel models, given the good internal consistency and inter-rater agreement of the ratings on the ten MQI whole-lesson items (see Chapter 3); the CFA model is used in the MSEM analyses of teaching and learning in the book and beyond. For a similar reason, we use the ISTOF measure the same way – a mean score for the multilevel models and the CFA model(s) for the MSEM analyses.

4.6 Masterly teaching through the qualitative lens

Key features of masterly teaching emerge in in-depth observations and interpretations of all lessons. These key features include, but are not limited to: (1) modelling the way for a shared discourse; (2) multiple representations for one mathematical fact; (3) not moving on until the class have reasoned in-depth the very essence of the task/topic; (4) variation as scaffolds for fundamental understanding, (5) lessons built on a variety of student contribution in an optimal sequence; (6) constant abstracting and generalising; (7) key structure of mathematics as the core of each lesson; (8) teacher gradually unfolding the essence of knowledge on the board; (9) eight other major features. As follows, we will first give a rich description of the

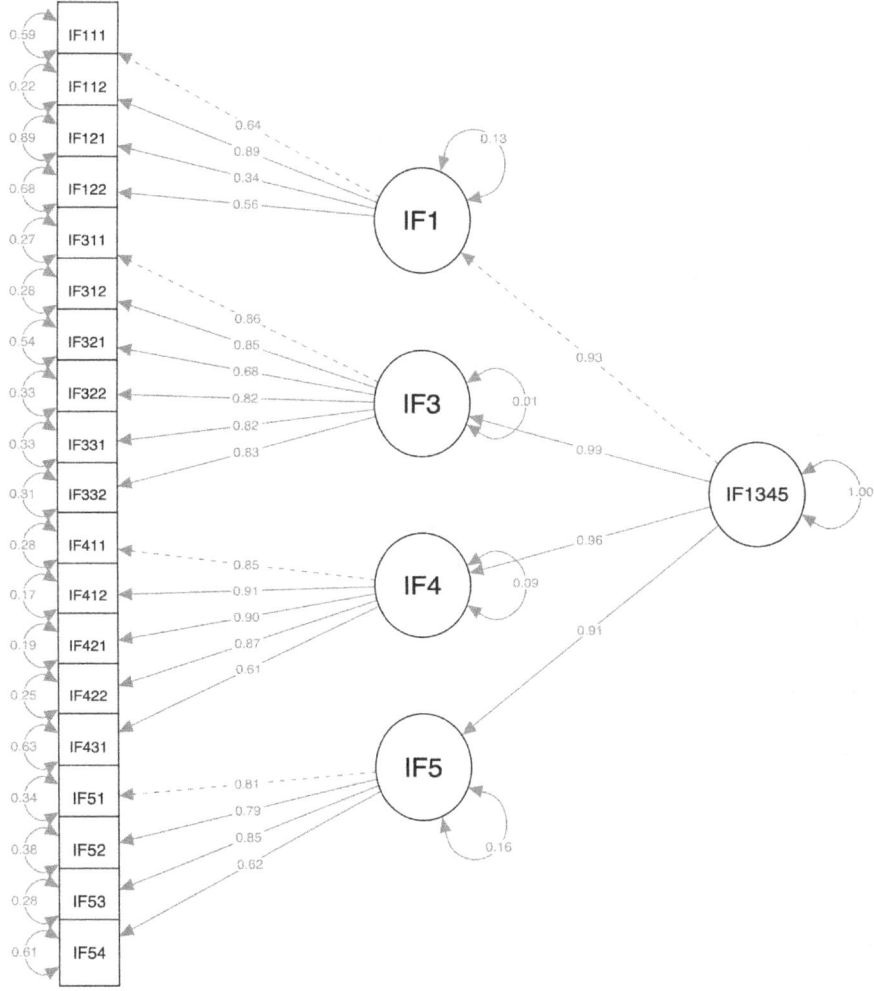

FIGURE 4.4   The second-order CFA model of ISTOF

former eight features with rich lesson examples and then discuss the latter eight features in-depth with no lesson examples.

4.7   Model the way for a shared discourse

Master teachers all clarify groupwork criteria or steps before students delve into the groupwork that might otherwise be less productive without clear goals and steps. It is with goals and standards clearly defined before independent/group work that fruitful outcomes can arise in the work of students and conceptual understanding can be built subsequently. Clear goals and standards are observable in all class-rooms. In many cases, necessary thinking 'tools' are offered so that students can use these to express their thinking without having to invent one themselves.

The topic in Ms S_{214}'s class is to solve real-world problems that can be modelled with geometric progression where the common ratio is 2. The lesson starts with the word problem shown on the screen about a teacher trying to get in touch and inform 15 choir members of an urgent show to take place. Before the class tap into the problem on their own, there is intensive class-wide brainstorming facilitated by the teacher on the key information and expectation of the problem and on the key strategies for solving it. After that, the teacher gives the class a final speedy formu-lation about the way in which they could represent their methods before they work independently on the task. She suggests the class use rectangles to represent the teacher in the problem and circles to represent the choir members, and label each line segment that connects a caller with a recipient with a number to represent the corresponding minute into the phone call. Having thoroughly analysed the problem within the entire class, students are now equipped with clear representing tools, ready for attempting the problem on their own. Such representing methods work as a unified language for the class to describe solutions for the same problem. It channels students' thinking towards the mathematical patterns behind the problem and its solution. Without it, more time might be lost with everybody struggling to anchor their thinking on a clear representation that must, initially, be sensible to themselves and, later, to their peers in the class.

Similarly, in Ms T_{416}'s lesson on revision of areas of 2D shapes, before group work on revision notes, the teacher asks the class to structure their discussion with peers and cover three aspects: (1) the name of the shape they have drawn; (2) its area formula; (3) how the formula is deducted.

In many of the classrooms, there are shared conventions for student-to-student interactions at the class level, in addition to the widely existing peer interactions during groupwork. Students seem to have gotten used to inviting peers to answer their questions or comment on their solutions during whole-class time, without having the teacher step in or wasting any lesson time on transitioning. These con-ventions have been long established and are now part of the class culture.

In Mr B_{303}'s lesson on problem solving, different students are invited to share with the class their solutions to two subtasks on finding best strategies for buying sportswear from a shop according to the specific requirements of the tasks. The second student, John, has written at the beginning of his solution a short paragraph in analysing the key information given in the task. This is applauded by the class, where applauding seems to be a class convention whenever they find their peers have done something great. After the applause, the teacher asks the class, 'What did you find in John's work?' This question leads to some of the students pointing out that John has given the task a clear analysis before solving it. The discourse convention we observed appears to be quite detailed, long-established and well carried out in the class.

4.8 Multiple representations, one mathematical fact

Another typical feature of masterly teaching is the promotion of multiple representations within the class to cultivate deeper understanding of the underlying mathematics. These representations almost all come from the students whose thinking is
motivated by chains of questions constantly posed by their teachers.

In Ms L_{304}'s class, the teacher started with a series of questions, responses and feedback about the rules of addition and subtraction of whole numbers. The conversation
smoothly moves to the rules of fraction addition and subtraction. Then, the discourse
naturally moves on to the addition of fractions with unlike denominators. Then she
posed the task $\dfrac{1}{2} + \dfrac{1}{4}$, asking the class to guess the answer. The remainder of the lesson proved that this task played a pivotal role in making learning happen and develop
ing cognition and knowledge in the learning brains. This is the core task around which
the fundamental knowledge of this entire lesson was embedded. This task has the
mission to transport the learner to the desired destination of knowledge. Upon initial
guesses in the class, two answers emerge, two-sixths and three-quarters. Two-sixths
was simplified as one-third, and later highlighted by a student who argued that one-
third is smaller than a half and could not be the sum of a half and a quarter which is
supposed to be something greater than a half. This left with the class one mission to
do – to explore and find out the answer and see if it's the hypothesised three-quarters.

After independent work, four students volunteered to write their solution processes on the board. As a result, five methods (Figure 4.5) were generated: two

FIGURE 4.5 Multiple representations of fraction addition created by students in
Ms L_{304}'s class

area models (#1 and #2); two arithmetic methods (#3 and #5) involving division and multiplication, and one method utilising the learnt method, reduction of fractions (#4). With the additional line segment method generated by a student (#6), the class have created, shared and discussed explicitly six methods for solving this one central task, $1/2 + 1/4$. These are impossible without the teacher's facilitation through intensive questioning before independent and group work.

4.9 Not moving on until the class have reasoned in-depth the very essence of the task/topic

A master teacher's class is always geared towards a thorough exploration of the mathematical laws and systems embedded underneath either real-world or mathematical phenomena. Such an exploration demands both independent and collaborative efforts. Collaboration happens not just at the group level but most importantly and largely at the class level. It is a whole-class exploration into the mathematical world.

At the heart of each lesson on new knowledge is one to two tasks, which feature(s) in almost all lessons. This feature reaches greater depth in master lessons. The purpose is for the entire class to learn the essential mathematics embedded in the problem. By reasoning about and solving this particular problem in greater depth, they are able to make generalisations and hence solve an entire category of problems.

For Ms S_{217}'s lesson titled *Making Phone Calls*, the underlying mathematics is *geometric progression*. In the textbook, *Mathematics 5B* (PEP, 2014, pp. 102–103), the task is set in a more general context of a school setting, whereas Ms S_{217} gives it a little 'twist' by setting it in the context of her own school. More specifically, it is about Grade 5 in the school which is exactly her students' year group (Figure 4.6). The lesson starts directly with the teacher presenting the word problem to the class. Then, the question-drive interaction takes place at the class level, with the teacher inviting the class to talk about the key information, the givens and the unknowns of the problem.

During the whole-class discussion of the key information in the problem, the teacher builds on student responses and points out the mathematical method of *Transforming Complex Tasks into Simple Ones* (化繁为简) that is widely accepted and practised in the Chinese classroom. She then sticks this four-character Chinese phrase on the right side of the chalkboard where teachers tend to put the key notes of a lesson. Turning back from the board, she suggests: 'So, what about we start with a smaller number, 7?' Then, the first task is split into two subtasks: the teacher informing seven students (task 1a); the teacher informing 432 students (1b). Task 1a lasts 29 minutes 5 seconds in which (1) the class explore the problem on their own first (4 minutes 22 seconds); (2) with the teacher posing questions, the whole class synthesise their solutions by sequentially looking at three students' plans that spend 5, 4, and 3 minutes respectively informing seven students (Figure 4.7); (3) the class work in groups to represent the best solution they agree upon in the last step and discuss this with their group members; (4) in the whole class, a volunteer

The textbook's version (PEP, 5B)	*Translation*
一个合唱队共有15人, 暑假期间有一个紧急演出, 老师需要尽快通知到每个队员, 如果用打电话的方式, 每分钟通知1人, 请帮助老师设计一个打电话的方案。	There are 15 members in a choir. During the summer break, an unexpected performance is needed. Teacher has to inform every member as soon as possible. If one person can be reached per minute by a phone call, could you help the teacher design a plan for the phone call?
Ms S_{217}'s version (task 1 or 1b)	*Translation*
我校五年级有432名学生, 暑假期间, 学校临时决定要组织五年级学生参加一项社会实践活动, 苏老师需要尽快通知到每个队员。如果用打电话的方式, 每分钟通知1人, 至少需要多少分钟呢？	There are 432 students in Grade 5. During the summer break, our school is going to organise a community immersion activity for the fifth graders. Ms Su needs to inform everyone as soon as possible. If one person can be reached per minute by a phone call, at least how many minutes are needed?

FIGURE 4.6 Ms S_{217}'s version of the problem on geometric progression

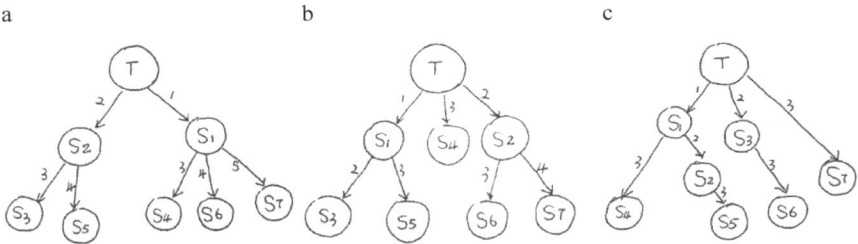

Student work samples redrawn by the first author of the book

FIGURE 4.7 Optimisation of problem solution built on student work in Ms S_{217}'s class

comes to the board upon invitation to represent the three-minute solution in a diagram step by step, whilst the teacher records the information in a table to the right of the diagram; (5) seeing the pattern gradually emerge, the class call out the total number of people receiving the notification, after which the teacher poses the crucial question, 'What if one more minute is given to the phone call?' This results in the teacher jotting down one more entry of information into the table (Figure 4.8a).

After this, task 1b is brought back. This part starts when the teacher quickly asks, 'As you all understand [the number of students we could inform in] four minutes, are you now able to solve this problem about informing 432 students in our year group?'

'Yes!' The class respond confidently and then go into independent work on the task.

Back to whole-class interaction, the teacher shows a student's work on the screen (Figure 4.8b) and asks the class about the pattern behind it. With the understanding built on task 1a over almost 30 minutes, the entire process of task 1b takes just 3 minutes 20 seconds, from the point when it is re-initiated to the point

a b c

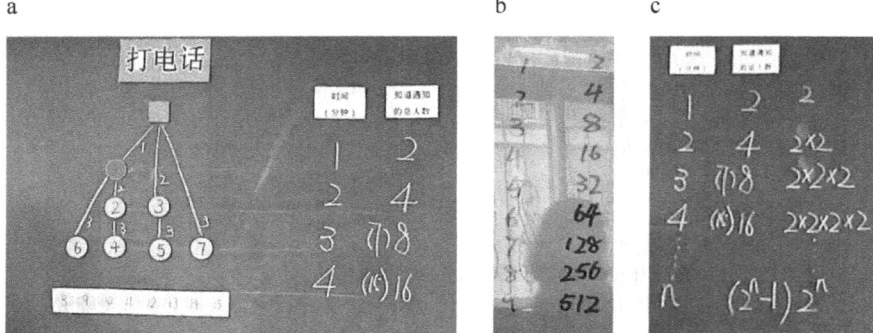

FIGURE 4.8 Pattern generalisation in Ms S_{217}'s class: from arithmetic to algebraic expression

when a whole class discussion about the solution is carried out. At the class level, the teacher guides the discussion towards completing the table on the right-hand side of the chalkboard by adding one more column to the right. This represents the mathematical model behind the problem and optimal solution (Figure 4.8c): the maximum number of people (teacher and students) having heard about the notification after n minutes into the phone call is 2^n, with ($2^n - 1$) being the maximum number of students having received the notification.

With the pattern generalisation and mathematisation of the real-world problem, students are now ready to model any given task of this type. Ms S_{217} seeks the opportunity to pose the second task: 'If the similar rule applies, what is the minimal number of minutes that is needed to inform the entire human beings (7.4 billion) on the planet of a message by phone?' This task is not to be completed by the class but acts as the extension of such a powerful pattern to real-world examples with a mega number. As the teacher gradually shows a specific number of minutes and the corresponding number of people receiving the notification, the class are astonished to see that as time shifts from 32 to 33 minutes, the number of recipients jumps from 4,294,967,296 to 8,589,934,592. Within 33 minutes, more than the current human population can be informed in the same way as task one defines.

The whole-class interaction on this task leads to the introduction of the formal name for the mathematics behind the problem, geometric progression, a specific category of compound interest. A quote from Einstein is shown on the next slide, regarding compound interest as the world's eighth wonder. After the quote, the teacher invites the class to give examples of geometric progression from the real world. A student gives the example about the chess inventor in ancient India who asks for the King to award him with the total of wheat placed in each of the 64 squares on the chessboard where the same growth pattern applies, from one square to the next. Then the teacher shows examples on the screen: the growth of duckweeds, the split of cells, the folding process of ramon and growth rates on the market.

4.10 Variation as scaffolds for fundamental understanding

Teaching with variation as a feature of Chinese mathematics teaching is widely accepted in practice and well documented in the research literature (Gu et al., 2004; Huang & Li, 2017).

Using three magnets on the board, Mr Z_{301} manages to develop amongst students a deeper understanding of the definition of triangles (segment 1) and a conceptual understanding of the height of a triangle (segment 2: Figure 4.9).

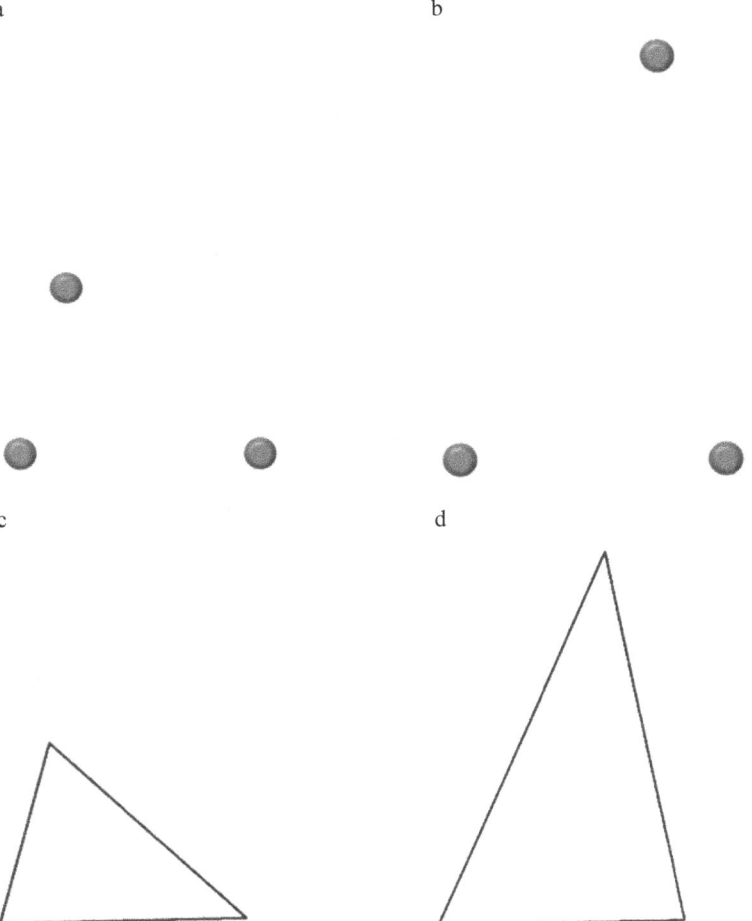

FIGURE 4.9 Developing the concept of the height of a triangle through variation in Mr Z_{301}'s class

We will have a closer look at segment 1 and then make a brief summary on segment 2. As the students come up with a unified definition of triangles, it is time to consolidate their understanding about it. Instead of emphasising the concept directly, the teacher processes this emphasis naturally over the course of attempting to represent a triangle with three magnet buttons on the board.

He asks the students to ignore the size of the magnets and imagine the three magnets as three points. Checking with the students that they can imagine the magnets as points by geometric standards, the teacher places two of the magnets (aligned horizontally) on the board and says, 'I am now placing the two magnets here as two vertices of a triangle. Where should this third one be placed so that they can be connected into a triangle?'

The majority of the class raise their hands, indicating they know how to do it and want to be invited to demonstrate it on the board. The teacher picks a boy who keeps his hands down and smiles. 'I know you must have ideas, though you haven't put up your hand.' The boy goes to the board and places the third magnet about a handspan above the magnet that is already placed on the bottom right by the teacher.

Mr Z$_{301}$:	Is that okay?
Class:	[clapping hands voluntarily] Yes.
Mr Z$_{301}$:	[indicating the boy to go back to seat whilst speaking to the class] Imagine it in your head. When it (that is, the third magnet) is placed here, what does the triangle look like)
Class:	[some attempting to describe it whilst moving their fingers in the air to draw the shape]
Mr Z$_{301}$:	[further encouraging all to describe the shape] Come on. Let's draw it by hands.

Then the teacher draws the triangle by hand following the class's instruction.

Mr Z$_{301}$:	Have you pictured it in your head?
Class:	Yes.
Mr Z$_{301}$:	Let me move it one more time. [moving the third magnet downwards to make the triangle like a pyramid being turned upside down] Is that acceptable?
Class:	Yes.
Mr Z$_{301}$:	Come on. Draw the triangle in gesture and show me where it is.
Class:	[each raising their hands and drawing the triangle in the air]
Mr Z$_{301}$:	[looking at the class thoughtfully] Okay, I'd like to ask you guys a question. If I give you three points, can you always connect them into a triangle?
Class:	[randomly answering; some raising hands wanting to share ideas] No . . . . That's not true . . . .

Mr Z₃₀₁:	Why on earth is it not true?
Class:	[some saying out loud]
	If you put them on a line, . . .
Mr Z₃₀₁:	[pretending not having a clue] Ah? Anne, come on. Show us your idea.
Anne:	[standing up and going to the board]
	If you put the three points on a line, ...
	[quickly moving the third magnet upwards and in between the other two magnets; turning to the class and continuing her explanation]
	they will connect into a line.
Class:	Right.
Mr Z₃₀₁:	Hang on a minute. What did she say? Placing the three points on the same line, right?
Class	Yeah.
Ms Z₃₀₁:	Placing them on the same line means that they cannot form a triangle, right?

With the class openly restating the condition 'three points must not stay on the same line', the teacher further introduces them to a formal term, *collinear*: 'In other words, the three points cannot be collinear.'

When the third vertex of a triangle is moved around, different triangles with the same base occur, but they are all triangles. However, when the third vertex falls in between the two vertices, they are on the same line, which means they are no longer vertices of a triangle. This way, the definition of a triangle is further consolidated at smaller details through variation.

In segment 2, also with the magnets, the height of a triangle emerges. As shown in Figure 4.9, by moving the third vertex higher from the former position, Mr Z_{301} asks the class what is the difference between the two triangles. The class naturally say it out loud that the latter is higher than the former – the height of a triangle as a concept is born!

In the above two segments of Mr Z_{301}'s lesson, the change of the third magnet's location serves two purposes: (1) deepening students' understanding of the definition of a triangle and (2) making the height of a triangle intuitive to the class before it is introduced (or in fact, this has become an introduction).

4.11   Lessons built on a variety of student contribution in an optimal sequence

Across classrooms, students' contributions play an important role in making the lesson tick. Without their contribution, teaching and learning is impossible. There is a wide range of contribution from students, including their work presented in various forms (correct or wrong) and the problems/tasks that they pose. The timing of their contribution is covertly and seamlessly arranged in optimal sequence by the teacher who has organised this behind the scenes with great care.

Rather than a burden to the teaching and learning momentum, student contribution is a powerful, if not sole, catalyser for learning. It is not contribution for contribution's sake. Learner response and learning results that constantly emerge in the lesson are instantaneously turned into teaching and learning resources, often through the constant use of 'peer reviewing' in the class regarding student answers and work samples.

As we continue to play back Mr Z_{301}'s lesson, one event soon arises, providing an example for student work as teaching and learning resources. After a conceptual understanding about the height of a triangle is established in the class, the teacher encourages everyone to go on and draw the height of a given triangle in the worksheet. Like many other classrooms, Mr Z_{301} takes the seatwork time to quickly 'cruise' the class and get a sense of the general trend amongst the students. He notices two typical answers: (1) drawing one height of the triangle; (2) drawing all three heights of the triangle (Figure 4.10). With both student presentation and class interaction driven by the teacher's questioning, the fact that a triangle has three heights is easily understood by everyone.

In many classes, creativity is nurtured through problem posing. We take Mr B_{303}'s lesson on *Problem Solving Strategies* for third graders as an example. The lesson is spread around six tasks, with the first two provided by the teacher and the other four posed by the students (Figure 4.11). Judging by the worksheet, you can only see three tasks, as the last task reads: 'the problem that I pose and my solution'. The last 23% of lesson time features four problems posed by four students, turning the seemingly 'last/third task' into four tasks. In the first 77% of the lesson time (Figure 4.11a), the class plan together and discuss their solutions for tasks 1 and 2 (Figure 4.11b), first with their desk-mates and then the whole class, using a quarter of the lesson time to look back at their problem-solving experiences and summarising lessons learnt and strategies accumulated. This gets them ready for posing thoughtful problems in the remaining 23% of the lesson time.

a
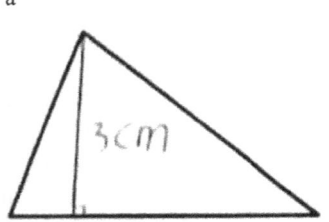

b
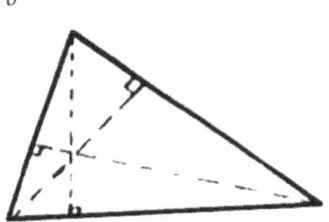

FIGURE 4.10 Triangle's height(s) presented and explained by Mr Z_{301}'s two students to the whole class via a document camera next to the board

a. Lesson structure

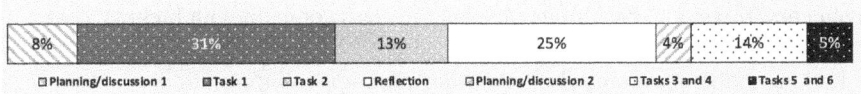

8%	31%	13%	25%	4%	14%	5%

| ☐ Planning/discussion 1 | ■ Task 1 | ☐ Task 2 | ☐ Reflection | ☐ Planning/discussion 2 | ☐ Tasks 3 and 4 | ■ Tasks 5 and 6 |

Note. The sections filled with dotted patterns are whole-class interactive activities.

b. Worksheet consisting of two given tasks and one problem-posing task

小明和爸爸带 300 元去运动服饰商店购物。

	85 元	16 元
130元	148元	108元	24 元

(1) 买一套运动服和一双运动鞋，最多剩下多少元？ [Task 1]
我的分析与解答：

我的提醒：
(2) 如果买两顶帽子，付出 100 元，最少找回多少元？ [Task 2]
我的分析与解答：

我的提醒：
(3) 我设计的问题与解答：[Original task 3]

Xiaoming and dad take 300 Yuan with them and go shopping at the sports store.
Pictures: sportswear prices in Chinese Yuan.
[Task 1]
To buy a tracksuit and a pair of shoes, what is the largest change that they can get in return?
My analysis and solution:
My reminder:
[Task 2]
To pay for two hats with a 100-Yuan note, what is the smallest change they can get in return?
My analysis and solution:
My reminder:
[Original Task 3]
The problem I designed and its solution:

c. Tasks posed by students under common given condition:

Xiaoming and dad take 300 Yuan with them and go shopping at the sports store.

[Tasks 3-4]

用300元去买一套运动服、一双运动鞋、一顶帽子，最多用去多少元？ 130<148 85<108 16<24
小明买一顶帽子，一套衣服，一双鞋，最多剩用多少元？

[Task 5]

要买一套运动服，一双运动鞋和一顶帽子共有几种不同的买法？

[Task 6]

店里所有衣服半价，买一整套服饰（包含衣服鞋子、帽子），最少找回几元？

[Task 3]
Taking 300 Yuan to purchase a tracksuit, a pair of shoes and a hat, what is the largest amount of money that they can spend?
130 < 148 85 < 108 16 < 24
[Task 4]
Xiaoming purchased a hat, a tracksuit and a pair of shoes, what would be the biggest change he could get in return?
[Task 5]
To purchase a tracksuit, a pair of shoes and a hat, how many kinds of shopping [combination] can they make?
[Task 6]
Everything in the store is 50% off. To purchase an entire set of sportswear (including a tracksuit, a pair of shoes and a hat), what would be the smallest possible change that they can get in return?

FIGURE 4.11 Tasks posed by the teacher and tasks posed by students in Mr B$_{303}$'s class

During the 77% of the lesson time on tasks 1 and 2, the whole class constantly reflect on the strategies for solving the given problems that are set under the same given conditions. The core strategies that the class come up with include:

- looking at the questions whilst thinking about given information;
- looking at the given information whilst thinking about the questions;
- to get the largest amount of change back means to buy the cheapest sportswear;
- to get the smallest change back means to buy the most expensive sportswear.

After a thorough reflection on the lessons learnt over two tasks provided by the teacher, the class go on to discuss in pairs the tasks that they have posed using the same given information. This is followed by the whole class solving four of the tasks that they posed. With the problem-solving and analysing strategies they obtained on tasks 1 and 2, the class are able to design meaningful problems that demand careful analyses, solutions and explanations. They are ready to make their thinking clear to peers in groups and then to the whole class.

In many of the revision lessons, students are invited to pose mathematical problems for the class. The sources of creation include solid understanding of the fundamental knowledge, previous mistakes they and their peers have made as well as their life experiences.

In Ms T_{416}'s class, students have completed the worksheet that their teacher gave them as homework. Today's revision lesson will be based on the worksheet. The task is to create a shape such that two perpendicular line segments, 4 cm and 2 cm long respectively, are part of the shape and that the shape's area can be calculated according to the two segments' lengths. The task expects the students to draw the shape and show the solution process for calculating its area. This is in fact a problem-posing task. Every student completing it may pose a unique version of the task on the area of a shape.

After a whole-class discussion on what is the purpose of revision in general and what is the expectation of the task, the class are asked to talk about their ideas and solutions in groups of four. Whilst they are chatting excitedly to each other, the teacher circulates the class to see everybody's work and gain an overview of the discussion. The main purpose is perhaps to pick the typical work samples. Near the end of the discussion, she invites nine students to draw their task figures and write the solutions on the board. They each have a different shape created, which makes the board a collection of a wide variety of 2D-shape problems on the same given information. There are nine tasks that stem from the initial problem-posing task on the worksheet. Each shape on the board is explained by its creator. Regarding each presentation, the teacher asks the class to recall the area formula of the specific shape which is shown using both the magnet card on the board and slide on the overhead screen.

The Lesson Segment 5 features Ms T_{416} presenting five problems posed by the students one at a time and interacting with the class about the solutions. These five problems are different from the initial task at the beginning of the lesson. They demand more thoughtful application of the properties of learnt 2D shapes and various knowledge related to measurement and shapes. The first three tasks are true or false (T or F) problems: (1) If two rectangles have the same perimeter, they also have the same area (F; not always so); (2) Two triangles of equal area can always form a parallelogram (F; only true when they are congruent); (3) For a square of side 4 cm, its area is equal to its perimeter (F; they are of different units, cm and cm^2). The fourth task asks: 'The length of a trapezoid's upper base is 5 cm, its lower base 4 dm, and its height 0.03 m. How many square metres is its area?' The fifth task asks for the length in centimetres of the base of a triangle with an area of 16 square centimetres and a height of 4 cm.

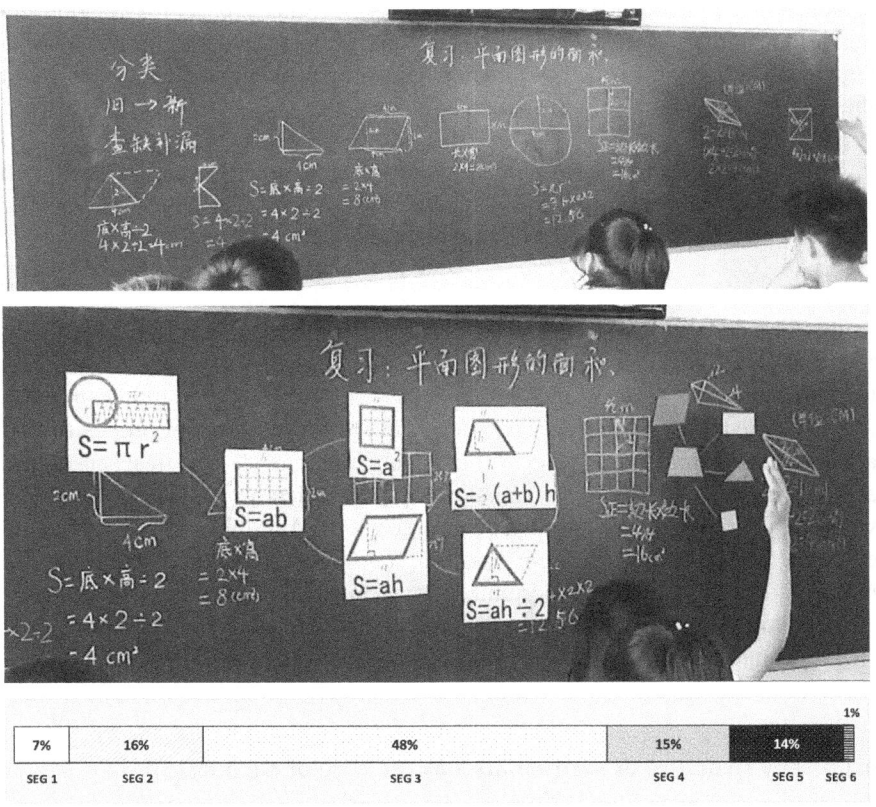

SEG 1: Whole-class discussion: purposes of revision in general and the task for discussion.   SEG 2: Group discussion;
SEG 3: Area formulas of learnt 2D shapes and proofs;   SEG 4: Conenctions between area formulas of 2D shapes;
SEG 5: Solving student-posed problems;   SEG 6: purpose of revision in general.

FIGURE 4.12   Tasks posed and knowledge structure mapped by the students in Ms T$_{416}$'s class

Students not only give their solution(s) but also their reasons for the solution(s), as well as the rationale and pitfalls that one must bear in mind for the particular type of problems. For example, task 4 seems easy but can be tricky, since the units for each piece of information given are different. It is hence suggested that caution should always be given to such details – otherwise, the problem would not be properly solved.

4.12 Constant abstracting and generalising

Throughout classrooms, there is constant abstracting of the real world to the mathematical world, over the course of which emerges mathematical models. However, the abstraction or modelling is performed with necessary stepping stones.

For students to solve the problems in an end-of-primary revision lesson on *Equality and Equation*, Mr W_{306} asks the class to first express the relationship between quantities in the form of a verbal equation and then in the corresponding algebraic equation, for which the unknown quantity is clearly defined. Take one of the six problems in a row in the lesson as an example:

> Three Gorges Dam has a volume 6.9 times as large as that of Liujia Dam, with a maximum capacity 336 *yi* (亿, Chinese place value for 0.1 billions) m³ greater than Liujia Dam. What is the approximate volume (*yi* m³) of Liujia Dam?
>
> (*PhEP, 2014d, p. 82: task 6 translated from Chinese*)

In the whole-class discussion after student independent work, the teacher asks, 'How do you express the quantitative relationship [verbally]?' A student volunteers to give the answer: ' *Volume of Three Gorges Dam – Volume of Liujia Dam* = 336 . ' Acknowledging the appropriateness of the answer, the teacher probes for the algebraic equation. Then, from those raising their hands, the teacher invites another student to express the relationship algebraically: $x - 6.9x = 336$. The verbal equalities thus act as a bridge between a word problem and an algebraic equation that models the problem situation. It connects real-world situations with mathematical models. The opportunity to systematically talk about these enables children to thoroughly understand the meaning behind abstract expressions consisting of numbers, letters and symbols, as they learn to model real-world problems algebraically.

4.13 Key structure of mathematics as the core of each lesson

The whole class, including the teacher, is actively building knowledge, with past, present and future learning in mind. In such collective exploration into the structured system of mathematics disappears the identity boundary between the teacher, who knows, and the learners, who presumably don't know yet.

A general pattern is that in a lesson when new knowledge is introduced for the first time, learnt components of the knowledge are recalled. The recalling process is often driven by teacher questioning. The purpose is to invite the past knowledge back and locate the new on the 'current' version of the mathematics map that is constantly growing as the class learn.

For instance, at the beginning of the lesson on *Dividing a Number by a Fraction*, Mr C_{204} asks the class to recall what they have learnt about fraction division. As the class give the immediate answer: Dividing a Fraction by a Whole Number, the teacher asks, 'What else should we learn but haven't?' Amongst the volunteers, a student is invited to respond, saying, 'Dividing a Fraction by a Fraction.' After receiving affirmation from the class on the student's answer, the teacher walks towards the chalkboard and writes to the right of the notes '*Fraction ÷ Whole Number*', the topic just pointed out by the student: *Fraction ÷ Fraction*.

Upon finishing the writing, he turns to the class with a smile, saying, 'This is what we haven't researched yet, right?' He pauses for a second and says, 'Some of you are still raising hands, aren't you?' Then, a boy is invited to speak. 'Dividing a Whole Number by a Fraction,' says the boy. 'That's right!' The class calls out loud. 'Ah? Have you thought about this?' speaks the teacher to the class in a gentle tone, 'Is that also fraction division?' 'Yes! Indeed!' respond the class immediately. Whilst writing '*Whole Number ÷ Fraction*' above the notes '*Fraction ÷ Fraction*', Mr C_{204} continues speaking: 'It could also be a whole number divided by a fraction.' After this, the teacher invites two more students to talk about their thoughts and they make their points clear in a super-concise fashion (45 seconds!).

Walking back to the board, Mr C_{204} points to two topics to the right and then back to the one topic to the left and asks, 'Look at those (right) and this (left). What is missing? Here.' In literally one second, the class call out, 'Dividing a Whole Number by a Whole Number!' Agreeing with the class, Mr C adds one more topic to the left of the first row, *Whole Number ÷ Whole Number*, a topic that has apparently been learnt long before the introduction of fractions.

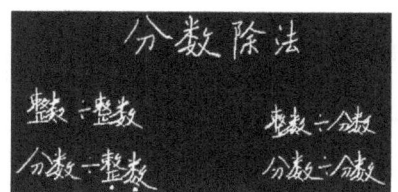

Mr C_{204}'s notes in Chinese

	Fraction Division	
Whole N ÷ Whole N		*Whole N ÷ Fraction*
Fraction ÷ Whole N		*Fraction ÷ Fraction*

(English Translation; Whole N = Whole Number)

FIGURE 4.13 The emerging structure of mathematics: Mr C_{204}'s notes on the board

4.14 Teacher gradually unfolding the essence of knowledge on the board

Traditionally, notes on the board in a Chinese classroom are one of the few key aspects of a lesson that everyone would look closely into, be it teachers, learners or an external observer/evaluator. These days, all the classrooms are equipped with smartboards which are normally inserted into the middle of the traditional board with two pieces of traditional board about the same size to each side. The title of a lesson is written on the board immediately when it has emerged in the class discourse. The main knowledge points appear on the board in the form of keywords when a sub-conclusion on a part of the lesson has just been reached collectively by the teacher and students. The smartboard is to complement the traditional chalkboard, but the chalkboard seems to still play the key role in exhibiting the essence of knowledge for each lesson. When the lesson comes to an end, it will have presented the completed map of knowledge points on the chalkboard. The smartboard shows either slides or problems/tasks for each section of the lesson, with the content changing from time to time, but the content on the chalkboard stays there resembling a constant force, the 'tree trunk' that holds together the leaves of the mathematics lesson. Everyone in the lesson, learning or teaching, knows the importance of it. It's the 'chalk way' of highlighting as one finishes another chapter of learning and teaching. Figures 2.10, 4.8, 4.12, 4.13 and 4.14 provide just a handful of examples.

4.15 Eight other features of the masterly teaching

Teaching is a system, instead of fragmented elements that are part of it. There are many other features of master teaching that we cannot cover with detailed lesson examples in this book. As follows, we attempt to name just eight more of them:

a. Knowledge map gradually unveiled on a slide of Ms W$_{311}$'s revision lesson on *Definition and Meaning of Fractions*

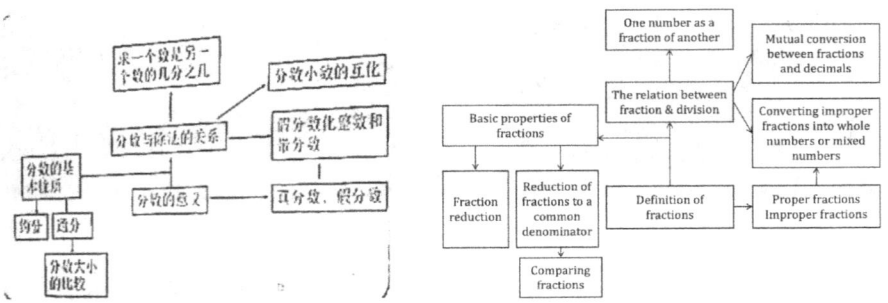

FIGURE 4.14   Knowledge maps in revision lessons: a few examples

b. knowledge maps of numbers (learnt in the primary stage) created by students in groups and by the teacher, Ms S₅₀₂, as a conclusion on knowledge mapping

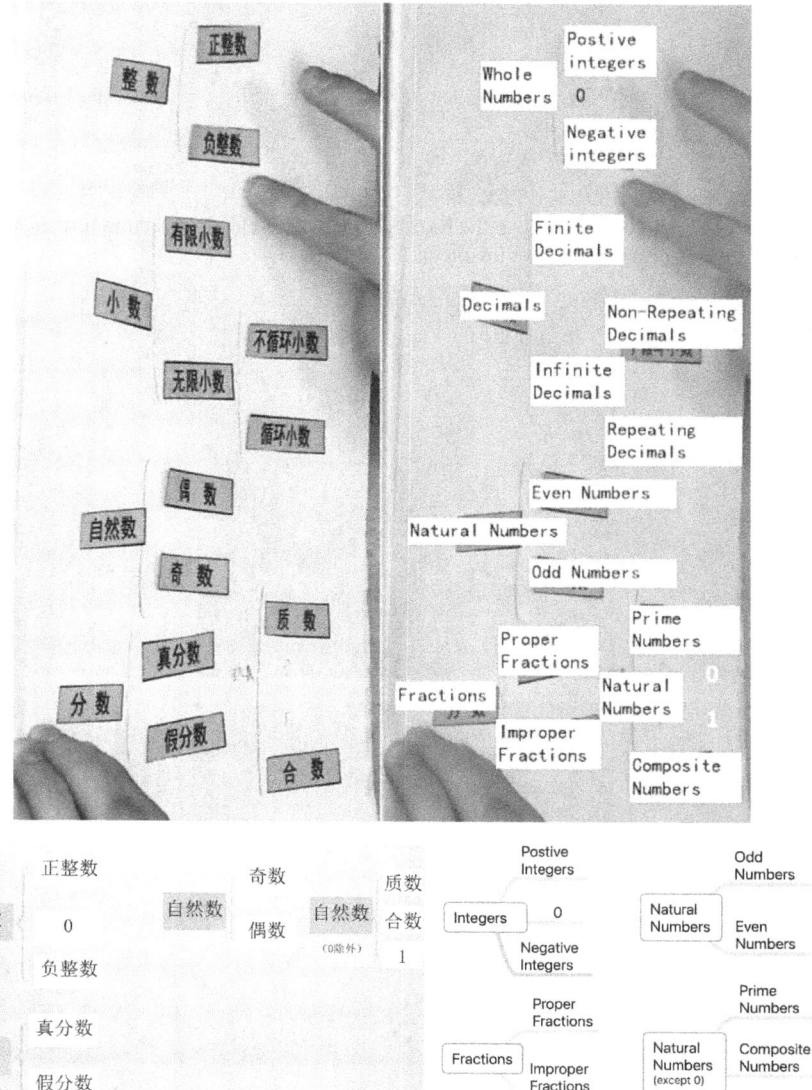

FIGURE 4.14   (Continued)

c. The expanding knowledge map in Ms L$_{304}$'s lesson on *Fraction Addition*

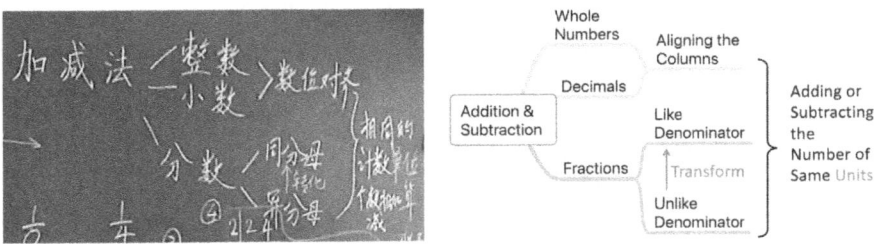

d. Knowledge map constructed on the board during whole-class interaction time in Mr L$_{309}$'s end-of-primary revision lesson on 2D shapes

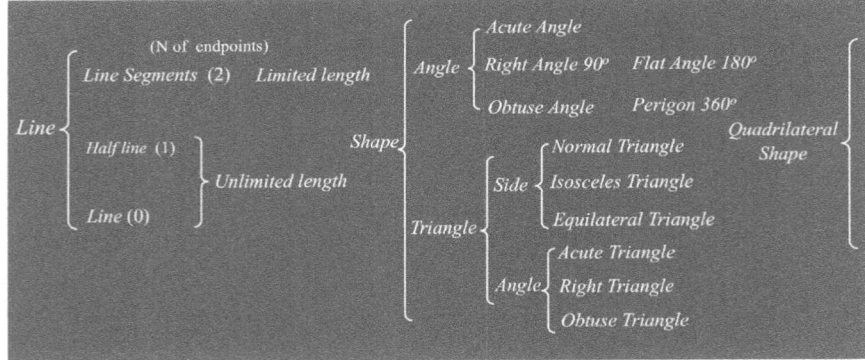

- As a common feature, master lessons are full of thoughtful teacher questioning, student responses and follow-up interactions. In fact, questioning (or not) doesn't matter. What matters is both the instant and ultimate results that culminate in the classroom. Most questions are not independent. They are often connected series that engage the learners in ongoing learning and enquiry into the mathematical content being taught. The series of questions are logically

e. Knowledge map on *Volume of 3D Shapes* gradually built by Ms Z_{314}'s class over the course of whole-class discussion and reflection on the content learnt and how each formula was derived when they first learnt it in the primary school

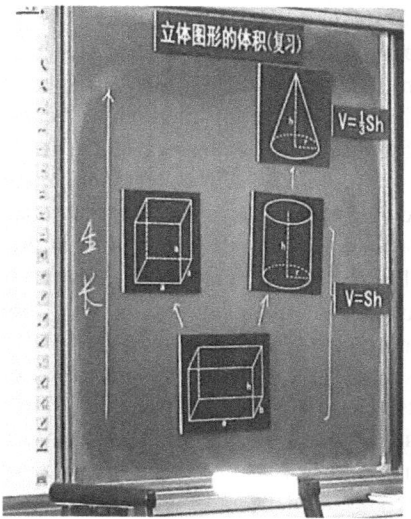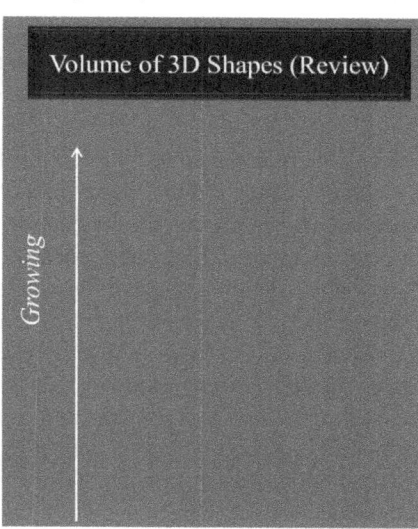

FIGURE 4.14 (Continued)

connected and cognitively progressive. Every step of practice is questioned and justified by the teacher and students. The questioning series does not shift direction until it reaches the bottom of the specific round of enquiry. In other words, *each series of questions leads to a cognitive milestone.*

- In addition to the process questions asking about the how and why of mathematics, teachers also ask lots of questions similar to the filling-in-blanks questions often seen in various tests. They tend to say a part of a statement upon which the class are building a conclusion and leave the *key words* for the class to say, such that the class and their teacher together complete the statement. The underlying assumption is that the mathematical truth is out there for the whole class, including the teacher, to discover through thinking, reasoning, doing mathematics and making summary on it *together.*
- Another important feature is that across classrooms, tasks are well-organised with a connected system of purposes. Tasks are organised in a *logical* order and *fit* well with the status of learners in the class. There are *core tasks* collectively and thoroughly tackled at the class level as well as *follow-up tasks* ready for students to try out new knowledge and skills in a super-efficient manner. The main tasks carry the mission for the class to generalise mathematical knowledge through both deductive and/or inductive reasoning. Ultimately in each lesson, it is not just one or several problems that the class aim to understand. It is an *entire category of problems* where essential knowledge, concepts or models embed. The tasks after the main tasks offer opportunities for an immediate transfer and timely enhancement of what is learnt on the main tasks.

- What masterly teaching aims for is deeper understanding, thinking and reasoning. The universally observable feature is *critical thinking in action*, with the entire class debating on key issues, such as correct and incorrect solutions to a problem, the rationale for using certain methods, the essence to a concept behind various mathematical phenomena, and the like. Even in those calculation lessons after the introduction lessons, practice is carried out through in-depth reasoning about the rationale behind the calculation method(s). Across lessons, there is a visible trend seeing learning move from *hands-on* to *heads-on* and ultimately from *written maths* to *mental maths*. Moreover, interwoven deep into the fabric of the question-driven classroom discourse is the practice of peer reviewing which has apparently become a part of the classroom culture. Every step is taken automatically with all peers joining in naturally. In revision lessons, there is a strong feature of reasoning from *abstract* to *concrete*, from *theory* to *practice*, from *inductive* reasoning to *deductive* reasoning, and vice versa. At the heart of each revision lesson are the outlines and properties of certain mathematics in a learnt unit, units or entire primary phase. The utmost emphasis is placed on the *meanings* of and *connections* between the otherwise fragmented pieces of mathematics, as the whole class reconstruct the *map of knowledge* independently and collectively (Figure 4.14 and Figure 2.1). There seems to be a shared map of mathematics that they constantly zoom in and out of to see both details and the big picture.
- Master teachers seek to facilitate deeper understanding of *not only the what but also the how and why*. Knowing mathematical facts is realised through understanding. Knowing the how is realised through understanding-based practice. Knowing the why is realised through reasoning, hypothesising, hypothesis-testing and debating. Beneath all this is the teacher-guided learning disguised, with good intentions, in the form of child-led inquiry into the unknown.
- One of the ultimate goals of masterly teaching is *teaching mathematics as a connected system*. In master lessons, there are multiple representations of mathematics, multiple solutions of problems and multiple peer-reviewing mini-events running throughout the masterly teaching we observed. Across classrooms, mathematics is never fragmented pieces of information – they are connected parts of the whole knowledge system. And mathematics is not just connected within itself. It is connected to the wider world, children's life experiences and individuals' responses to questions or problems arising in that particular lesson. This is observable across lessons where a topic is being introduced, practised or reviewed. Students are constantly reminded to seek connections, such that it has almost become a mindset – the *connecting mindset* that does not just see knowledge as loose dots but interconnected growing systems. However, the masterly teaching attends to both small details and big ideas in mathematics. It is by constructing each chunk of knowledge in great depth that they can be gradually added to the knowledge map and remain permanently in children's minds. The knowledge building works in cycles. It starts from limiting the mind's focus to

small 'leaves' of knowledge at greater depth and then temporarily ends on adding the newly grown leaves onto the 'knowledge tree'. A new cycle like this is initiated when a new topic is introduced. The process builds on itself.

- The major mission of the teacher and students in a master lesson is *thinking and reasoning about fundamental meaning of mathematics*. Throughout the classrooms, teachers pose questions to challenge students to justify the solutions or hypotheses they or their peers make. Their reasoning is made audible and visible by the opportunities that their teachers give them in the group- and class-level discussion sessions. They are encouraged to make connections between mathematical phenomena and concepts, between mathematics and the real world, and between their ideas and their peers' ideas and so forth. Errors become precious thinking, reasoning and natural learning opportunities in every classroom. Some classrooms even have the convention that oversees students interact and debate with each other upon their understanding and solutions in a very professional and efficient manner.

- Independent and collaborative learning is cultivated at the same time. An observable shared value across classrooms is that one should be able to think through and solve problems independently whilst maintaining an excellent role in working collectively with peers. It is important that each student is expected to think about and do mathematics independently before they join each other in groups or the whole class to think collectively and critically. The collaboration is more nurtured during the whole-class interactive time than groupwork. The entire class – students and the teacher – go through a thorough exploration into the focal mathematics and its connections with the wider world or vice versa, during the whole-class interactive time. The exploration is coherent with a clear boundary, putting both the children and mathematics at the centre.

4.16   Master lessons: so, what?

We have been to master teachers' classes and seen the lessons through both quantitative and qualitative lenses. As the Integrated paradigm seeks to capture, we have seen both sides of the coin or – more precisely – multiple dimensions and multiple aspects of the masterly teaching reality. With the master lessons in mind, let us travel further to the following three chapters to see the three types of mathematics learning outcomes and the corresponding teaching-learning mechanisms that significantly explain the educational excellence. First up, the affective outcomes.

5

AFFECTIVE OUTCOMES

Patterns, variation and mechanisms

Chapter overview

- Measuring and modelling affective outcomes
- Do students feel they belong?
- Does the mathematics teaching engage students?
- Do students like mathematics?
- General patterns, variation and correlation of student perceptions
- Teaching that makes mathematics enjoyable
- The paths towards better affective learning outcomes in maths
- Patterns, variation and mechanisms of affective learning outcomes in maths

> Education breeds confidence. Confidence breeds hope. Hope breeds peace.
> — Confucius

> I always liked numbers.
> — Terence Tao

Education leaves students with an accumulated impression of schooling, teaching and learning, amongst other things. These together reflect both the quality and outcomes of education they obtain.

Belonging is a basic human need as a social creature, and *school belonging* (SBL) is a basic need for those attending schools. Having a sense of school belonging means that students have positive relationships with their peers, teachers and other adults in the school, feeling accepted, confident, happy, safe and worthwhile being there (Allen et al., 2018; Slaten et al., 2016). A strong sense of

DOI: 10.4324/9781003127925-5

school belonging not only serves as a desired consequence of schooling but also contributes to a positive school climate under which children become academically engaged (Neel & Fuligni, 2013; Thapa et al., 2013) and well developed in executive functioning which contributes to their success in learning and life (Cumming et al., 2020). SBL can thus be seen as both the cause and result in and of itself. As a latent concept, it is often inferred by student self-reported measure on a Likert scale (e.g., the one used by IEA, 2014). In a sense, it is also learners' evaluation of the quality of schooling, as in the case of student-perceived teaching quality (Scherer & Gustafsson, 2015).

Students are the direct observers (Kyriakides et al., 2014) witnessing whether their mathematics teachers' teaching can engage them in the kind of learning that makes sense to them. The way students perceive teaching, if measured reliably and validly, can largely reflect the quality of teaching (Bijlsma et al., 2022; Kyriakides et al., 2014; Scherer & Gustafsson, 2015). Good teaching is accurate, clear and well-structured; it adapts to learning difficulties, facilitates active learning and promotes metacognitive thinking; it formulates a positive learning environment (Muijs et al., 2014; Teddlie et al., 2006). Good mathematics teaching conveys deep and rich mathematics, representations and solutions; it promotes mathematical thinking, reflection and communication; with it, student work is utilised as invaluable learning and teaching resources (Learning Mathematics for Teaching, 2011).

Among all the aims in mathematics education, an important aim is to seed the joy of mathematics learning (a.k.a. affective outcomes) amongst learners. One way in which the mathematics education literature describes affective aspects of mathematics education is through the term attitudes towards mathematics (ATM). The multi-faceted nature of ATM gives rise to many definitions of the construct (e.g., Di Martino & Zan, 2010; Hannula, 2002; Kiwanuka et al., 2017). Hart (1989) defines ATM in terms of three dimensions: an affective, a cognitive and a conative (behavioural) dimension, hence referring to a combination of the emotions that students associate with mathematics, their beliefs towards mathematics, which could be either positive or negative, and how they behave regarding mathematics. Most definitions refer to the interplay of several elements. Considering both student and classroom influences on ATM, Kiwanuka et al. (2017) define ATM as 'an aggregated measure of mathematics self-confidence, perceived usefulness, and enjoyment of mathematics' (p. 3).

Numerous factors influence ATM. Students' ATM can be influenced by parental education levels and occupations (Köğce et al., 2009; Mohamed & Waheed, 2011) as well as parents' own ATM (Beswick, 2006; Schoenfeld, 1989). Mixed results exist on gender differences in ATM, with some studies showing no significant difference in ATM between male and female students (e.g., Köğce et al., 2009; Mohd et al., 2011), and other studies showing significant differences (e.g., Frost et al., 1994). Studies on the relationship between ATM and achievement in mathematics (AIM) have also resulted in mixed results, with some indicating a reciprocal relationship (Ma & Xu, 2004; Minato & Kamada, 1996) but some a unidirectional relationship (e.g., Garon-Carrier et al., 2016). About the ATM-AIM relationship, the meta-analysis of 113 studies by Ma and Kishor (1997) suggests an overall positive effect, with a weaker association in primary schools and a stronger relation in secondary schools.

As part of the mathematics survey, we asked students about their sense of school belonging (SBL), views on the engagement of mathematics teaching (EGM) and ATM. To measure SBL, EGM and ATM, we utilised three scales from the TIMSS 2015 background questionnaire for Grade 4 (IEA, 2014), with reuse permission approved by the IEA (#IEA-19–008). The former two scales, SBL and EGM, are essentially about the quality of schooling and teaching from the students' points of view. The ATM scale measures affective learning outcomes in mathematics – whether or not students like mathematics as a result of learning it. We consider SBL and EGM play two roles in our data analyses: (1) general affective outcomes of schooling and teaching; (2) predictors of affective learning outcomes in mathematics – ATM. Therefore, we may call all the three variables affective outcomes in a broad sense, whilst regarding ATM as the ultimate affective learning outcomes in mathematics. In the teaching-learning models that we run in the later part of the chapter, we treat SBL and EGM as predictors of ATM, scrutinising the schooling-teaching-learning loop in the affective dimension at the student and teacher levels.

In the remainder of the chapter, we first explain the methods for measuring SBL, EGM and ATM and the techniques for modelling the potential relationships between them. For each of the three constructs, we present the results of a CFA and explain the fit of the latent variable model to the data. To show the general patterns of affective outcomes, we give the descriptive statistics on each construct and explore the primary connections between them by placing them in a correlation matrix. Having the general patterns in mind, we go on testing the following two hypotheses using multilevel modelling and multilevel structural equation modelling:

H_{AF1}: There are significant teacher-level effects on student attitudes towards mathematic.

H_{AF2}: Student sense of school belonging affects significantly their perceived teaching engagement which in turn affects their attitudes towards mathematics.

5.1 Measuring and modelling affective outcomes

As introduced at the beginning of the chapter, three types of education outcomes were measured with three scales from the TIMSS 2015 questionnaire for Grade 4 (IEA, 2014). These Likert scales measured: (1) SBL (seven items: ASBG11A-G), (2) EGM (nine items: ASBM02A-J), and (3) ATM (ten items: ASBM01A-I). For each scale, the response ranged from 0 (disagree a lot) to 3 (agree a lot). Items 2 and 3 in the ATM scale were stated negatively, hence being reverse-coded.

The general patterns and variations of the three measures as item parcels (mean score) are explored through descriptive statistics, t-tests (variation by gender), zero-order pairwise correlations between the constructs and between each of them and student age or SES.

The structure of the three latent variables (SBL, EGM and ATM) is each explored with an explorative factor analysis and verified with a CFA. In modelling the teaching-learning mechanisms, we use the item parcels (mean score) of these constructs as variables in multilevel models and each construct's CFA model as a measurement part of structural equation models.

5.2 Do students feel they belong?

Overall, on a scale of 3, each of the SBL items receives an average rating of above 2.5 (see Table 5.1), with the ratings of all seven items averaged 2.64 (SD = 0.46). The item with the highest average rating and the smallest SD asks if students feel proud attending their schools (SBL6). In their schools, students feel that they belong (SBL3) and that their teachers (SBL5) and fellow students (SBL4) are nice to them. They enjoy going to school (SBL1), learn a lot there (SBL7) and feel safe at school (SBL2).

An α value of 0.82 suggests that the SBL scale has a good internal reliability across its seven items. The CFA results further indicate that the scale as a latent variable model fits well with the data (χ^2 = 150.709, df = 14, p < 0.001, CFI = 0.976, TLI = 0.964, RMSEA = 0.058, SRMR = 0.025). The path coefficients of the seven items range from 0.55 to 0.70 (p < 0.001), which indicates strong correlations of the manifesting items with their underlying concept – SBL.

5.3 Does the mathematics teaching engage students?

On the EGM scale (Table 5.2), the highest average rating with the smallest standard deviation is given to the quality of teachers' explanation (EGM6). Students understand that their teachers maintain a rich level of teaching (EGM8) and are good at

TABLE 5.1 Descriptive statistics and the CFA factor loadings for SBL

Items	Item information	N	Mean	SD	CFA factor loadings
SBL1	Happy at school	2950	2.56	0.72	0.70***
SBL2	Safe at school	2945	2.52	0.76	0.64***
SBL3	Sense of belonging	2951	2.71	0.61	0.70***
SBL4	Positive peer relation	2945	2.57	0.75	0.55***
SBL5	Fairly treated by teachers	2947	2.66	0.66	0.57***
SBL6	Feeling proud	2944	2.83	0.48	0.68***
SBL7	Learning a lot	2950	2.56	0.72	0.60***

Note: Item source: (IEA, 2014). *** p < 0.001.

TABLE 5.2 Descriptive statistics and the CFA factor loadings for EGM

Items	Item information	N	Mean	SD	CFA factor loadings
EGM1	Clear teacher expectations	2874	2.46	0.69	0.56***
EGM2	Easy to understand	2868	2.67	0.59	0.59***
EGM3	Interested in teacher speaking	2870	2.65	0.63	0.74***
EGM4	Interested in tasks given	2873	2.33	0.84	0.68***
EGM5	Clear answers	2873	2.75	0.55	0.65***
EGM6	Clear explanation	2873	2.87	0.42	0.59***
EGM7	Opportunity to show work	2868	2.54	0.70	0.63***
EGM8	Rich content to engage students	2866	2.82	0.49	0.68***
EGM9	Using student misconception	2868	2.79	0.52	0.70***
EGM10	Willing to listen to students	2867	2.58	0.71	0.63***

Note. Item source: (IEA, 2014). ***$p < 0.001$.

building teaching on learners' misconceptions (EGM9). With good teaching clarity, teachers respond to students with clear answers (EGM5) and communicate with students in an understandable manner (EGM2). Students thus enjoy listening to their teachers (EGM3) who are, in turn, good listeners to students (EGM10). Student work becomes teaching and learning material as teachers offer students opportunities to present their work in the class (EGM7). Teaching with clarity also means that clear expectation is channelled towards students (EGM1). Teaching becomes engaging when the tasks given to students spark interest in them (EGM4).

A good internal reliability is found in the EGM scale ($\alpha = 0.870$). As indicated by the CFA results, the latent variable model of EGM fits adequately to the data ($\chi^2 = 820.425$, $df = 35$, $p < 0.001$, CFI = 0.924, TLI = 0.902, RMSEA = 0.089, SRMR = 0.044). The large sample size might have led to the insignificance of the Chi-square test. Strong bivariate correlations are observed between the latent variable, EGM, and each of its ten items, with the correlation coefficients ranging from 0.56 to 0.74 ($p < 0.001$).

5.4 Do students like mathematics?

On a scale of 3, students give an average rating greater than 2.3 on all the nine ATM items. As shown in Table 5.3, the highest average rating falls on the statement that they learn lots of interesting things in mathematics (ATM4). Three items on liking mathematics (ATM2 and 5) or enjoying mathematics learning (ATM1) each have a mean rating greater than 2.6. The rest of the scale see students give an average rating of 2.51, 2.44 and 2.38 respectively on: mathematics is not boring, mathematics is a favourite subject and longing for mathematics lessons (ATM3, 9 and 8). The remaining items get an average rating of 2.43 and 2.33 on: the enjoyment of solving mathematics problems (ATM6) and mathematics schoolwork (ATM7) respectively.

TABLE 5.3 Descriptive statistics and the CFA factor loadings for ATM

Items	Item information	N	Mean	SD	CFA factor loadings
ATM1	Like learning maths	2943	2.65	0.62	0.83***
ATM2REV	Dislike learning maths	2934	2.67	0.71	0.54***
ATM3REV	Maths is boring	2933	2.51	0.79	0.59***
ATM4	Maths is interesting	2938	2.73	0.58	0.69***
ATM5	Like maths	2939	2.62	0.67	0.87***
ATM6	Like maths schoolwork	2936	2.33	0.80	0.73***
ATM7	Like solving maths problems	2941	2.43	0.77	0.76***
ATM8	Like maths lessons	2939	2.38	0.79	0.83***
ATM9	Maths is favourite subject	2942	2.44	0.84	0.82***

Note: Item source: (IEA, 2014). REV: Reversely coded. ***$p < 0.001$.

With an α of 0.914, the nine items have a sound internal reliability. Apart from the significant chi-square test due to the large sample size and a marginal RMSEA, the CFA results indicate an acceptable fit of the latent variable model, ATM, to the data ($\chi^2 = 1043.921$, $df = 27$, $p < 0.001$, CFI = 0.937, TLI = 0.916, RMSEA = 0.114, SRMR = 0.043). There are significant and strong correlations between each of the nine items and the underlying concept, ATM, given that the coefficients range from 0.54 to 0.87 ($p < 0.001$).

5.5 General patterns, variation and correlation of student perceptions

On a scale of 3, the cohort of students score relatively high on SBL, EGM and ATM, with the mean score of each measure being 2.5 or higher. The *t*-test results (Table 5.4) indicate that girls rate SBL and EGM significantly – though not substantially – higher than boys who, however, score slightly higher on ATM than girls. In the correlation matrix, a similar pattern occurs between gender and the three constructs, when gender is treated as a dummy variable (0 = boy, 1 = girl). Judging by the zero-order correlational results, age and SES had little effect on any of the three measures. EGM was significantly strongly related to both SBL and ATM which were mutually related with a significant coefficient approaching the cut-off point for a strong effect.

As discussed at the beginning of the chapter, we see ATM as the ultimate affective outcome for the learning of mathematics. Now we are ready to look for factors that contribute to positive attitudes towards mathematics. In the following two sections, we study the teaching-learning mechanisms that lead to such attitudes, through multilevel modelling and multilevel structural equation modelling.

TABLE 5.4 Variable description, *t*-tests and correlation matrix

Variables	Mean	SD	Min	Max	Age	Gender	SES	SBL	EGM
Age	11.6	0.6	9.4	13.6					
Gender	0.5	0.5	0	1					
SES	34.2	7.4	9.2	51.3	-0.04*	0.02			
SBL	2.6	0.5	0.0	3.0	0.02	0.12**	0.02		
EGM	2.6	0.4	0.0	3.0	0.04*	0.05**	0.00	0.60**	
ATM	2.5	0.6	0.0	3.0	-0.02	-0.08**	-0.02	0.47**	0.68**

t-test	*Girl* M (SD)	*Boy* M (SD)	*t*	*df*	*p*	*d*
SBL	2.694 (0.403)	2.591 (0.502)	6.213	2934	***	0.229
EGM	2.669 (0.409)	2.623 (0.439)	2.939	2859	**	0.110
ATM	2.485 (0.594)	2.566 (0.537)	3.900	2777	**	0.145

Note: SBL, EGM and ATM here are item parcels (mean scores). *$p < 0.05$. ** $p < 0.01$.***$p < 0.001$

5.6 Teaching that makes mathematics enjoyable

To test the hypothesis that teaching affects ATM (H_{AF1}), we planned to run three multilevel models: ATM_{MLM0}, ATM_{MLM1} and ATM_{MLM2}. First of all, to test whether the structure of data warranted multilevel analyses, we ran a null model (ATM_{MLM0}) which resulted in an intraclass correlation (ICC) value of 0.12. This means that 12% of the variance in ATM is located at the teacher level and hence justifies the existence of a multilevel structure in the ATM data. We therefore proceeded to run the other two models. In order to examine the possible effects of background variables, we added the three control variables (age, gender and SES) to the null model and got the second model (ATM_{MLM1}). The results showed a very small drop in the ICC (0.11), a reduction of the ATM variance by 2.4% at the student level but an increase of the variance at the teacher level by 9.3%.

To test the hypothesised effect of teaching on the student ATM, we added the teaching variable, the aggregated EGM (L2EGM), to the former model (ATM_{MLM1}) and got the full model, ATM_{MLM2}. The following equations describe the full model:

Level-1 equation:

$$ATM_{ij} = \beta_{0j} + \beta_{1j}Age_{ij} + \beta_{2j}Gender_{ij} + \beta_{3j}SES_{ij} + e_{ij},$$

Level-2 equations:

$$\beta_{0j} = \gamma_{00} + \gamma_{01}L2EGM_j + u_{0j},$$
$$\beta_{1j} = \gamma_{10},$$
$$\beta_{2j} = \gamma_{20},$$
$$\beta_{3j} = \gamma_{30}.$$

Combined equation:

$$ATM_{ij} = \gamma_{00} + \gamma_{10} Age_{ij} + \gamma_{20} Gender_{ij} + \gamma_{30} SES_{ij} + \gamma_{01} L2EGM_j + u_{0j} + e_{ij}$$

In the equations, ATM_{ij}, Age_{ij}, $Gender_{ij}$ and SES_{ij} are the ATM score, age, gender and SES of student i in class j respectively; $L2EGM_j$ is the aggregated EGM in class j; e_{ij} and u_{0j} are the residual error terms at levels 1 and 2 respectively.

As shown in Table 5.5, the results of the full model (ATM_{MLM2}) show a reduction of the ATM variance from the null model by 96.7% at the teacher/class level and by 2.2% at the student level. Comparing the models ATM_{MLM1} and ATM_{MLM2}, the latter apparently works better in explaining the variance of ATM. In the former, level-2 variance does not decrease but increases instead, after adding the three background variables to the null model. In the latter and final model, the aggregated EGM explains almost all the variance at the class level, even when age, gender and SES are kept as control variables.

In summary, the teaching variable explains the affective learning outcomes in mathematics dramatically: there is 12% of the ATM variance situated at the teacher/class level; 96.7% of such variance is explained by the aggregated EGM. Teaching does make a difference in affective outcomes of mathematics, and teaching engagement can indeed cultivate positive attitudes towards mathematics amongst students.

TABLE 5.5 Two-level models of ATM regressed on aggregated EGM

Model	ATM_{MLM0}		ATM_{MLM1}		ATM_{MLM2}	
Fixed part	Coefficients	ES	Coefficients	ES	Coefficients	ES
(Intercept)	2.524***		2.542***		0.038	
Student level						
Age			-0.008		-0.030**	-0.20
Gender			-0.092***	-0.60	-0.095***	-0.62
SES			0.004*	0.03	0.002	
Teacher level						
L2EGM					1.065***	6.92
Random part	Variance	SD	Variance	SD	Variance	SD
Student level (σ^2)	0.303	0.550	0.296	0.544	0.296	0.544
Teacher level (τ_{00})	0.024	0.154	0.026	0.161	0.001	0.028
ICC	0.12		0.11		0.05	
$N_{Teacher}$	67		67		67	
$N_{Observation}$	2947		2813		2813	
Deviance	4941.6		4657.3		4565.8	
Marg R^2/Cond R^2	0.000/0.072		0.008/0.088		0.079/0.081	
Variance Reduction from Null (%)						
Student level			2.4%		2.2%	
Teacher level			-9.3%		96.7%	

Note: $ES = \dfrac{\text{coefficient}}{SD_{\tau_{00}, NullModel}}$. EGM = student-perceived teaching engagement in mathematics class measured with the scale ASBM02A-J in TIMSS 2015 Student Questionnaire for Grade 4 Mathematics (IEA, 2014). * $p < 0.05$. ** $p < 0.01$. *** $p < 0.001$.

5.7 The paths towards better affective learning outcomes in maths

In this section, we ran a series of four structural equation models to test the hypothesis that students' sense of school belonging (SBL) affects their perceived engagement of mathematics teaching (EGM) which in turn affects affective learning outcomes in mathematics – ATM (H_{AF2}). Models are run with the R package, *lavaan*, in R and RStudio (R Core Team, 2021; Rosseel, 2012; RStudio Team, 2021). The first two models (ATM_{SEM0} and ATM_{SEM1}) were single-level SEMs; the other two were multilevel SEMs (ATM_{MSEM0} and ATM_{MSEM1}).

As shown in Table 5.6, all four models have an acceptable fit to the data, with the last model (ATM_{MSEM1}) generating the smallest value of χ^2 / df. Comparing the two MSEMs, though the model ATM_{MSEM0} has a smaller value of Akaike's information criterion (AIC) or Bayesian information criterion (BIC), the last model takes into consideration the three control variables (age, gender and SES). We therefore deem the model ATM_{MSEM1} better. The path coefficients of the model (Table 5.7) show that SBL positively affects EGM at both the student and the teacher levels $\left(\beta_{L1} = 0.65, p < 0.001; \beta_{L2} = 0.89, p < 0.001\right)$ and that ATM is positively affected by EGM at both levels $\left(\beta_{L1} = 0.67, p < 0.001; \beta_{L2} = 1.54, p < 0.01\right)$ as well, with the direct effect of SBL on ATM being significant and weak at the student level $\left(\beta = 0.10, p < 0.001\right)$ but not the teacher level.

TABLE 5.6 Model fit indices for structural equation models of ATM

SEM/MSEM: SBL→EGM→ATM

Model	χ^2		df	p	χ^2/df	CFI	TLI	RMSEA
ATM_{SEM0}	3394.538		296	0.000	11.468	0.911	0.903	0.062
ATM_{SEM1}	3501.364		368	0.000	9.515	0.906	0.897	0.057
ATM_{MSEM0}	3590.342		592	0.000	6.065	0.911	0.902	0.043
ATM_{MSEM1}	3673.126		736	0.000	4.991	0.908	0.900	0.039

Model	$SRMR_w$	$SRMR_b$	AIC	BIC	N	Note
ATM_{SEM0}	0.045		108838.175	109163.840	2755	1-level without controls
ATM_{SEM1}	0.044		103459.856	103818.308	2634	1-level + Age, Gender & SES
ATM_{MSEM0}	0.046	0.090	108332.200	109137.480	2755	2-level without controls
ATM_{MSEM1}	0.043	0.114	125910.475	126780.161	2634	2-level + Age + Gender + SES

Note: ATM_{SEM0} and ATM_{MSEM0} are models without controls; ATM_{SEM1} and ATM_{MSEM1} are models controlling for age, gender and SES. CFI = comparative fit index. TLI = Tucker-Lewis index. RMSEA = root mean square of approximation. SRMR = standardised root mean square residual.

TABLE 5.7 Path coefficients of structural equation models of ATM

Model	Student level			Teacher/Class level		
	EGM~SBL	ATM~EGM	ATM~SBL	EGM~SBL	ATM~EGM	ATM~SBL
ATM_{SEM0}	0.69***	0.72***	0.05*			
ATM_{SEM1}	0.68***	0.70***	0.08*			
ATM_{MSEM0}	0.66***	0.69***	0.07*	0.91***	1.47***	-0.57
ATM_{MSEM1}	0.65***	0.67***	0.10***	0.89***	1.54**	-0.66

Note: $*p < 0.05$. $** p < 0.01$. $*** p < 0.001$.

5.8 Patterns, variation and mechanisms of affective learning outcomes in maths

In this chapter, we have looked at the affective outcomes of schooling and mathematics teaching and learning. A general pattern is that the students from master mathematics teachers' classrooms give quite high ratings about school belonging, teaching engagement and the enjoyment of mathematics learning.

Whilst children's age and SES have little influence on their sense of schooling, perceived teaching quality and attitudes towards mathematics, they do differ in these perceptions by gender. Though on average both gender groups rate SBL, EGM and ATM at quite a high value, girls perceive the quality of schooling and teaching slightly more positively than do boys who, on the contrary, find mathematics a little more enjoyable than girls.

Multilevel modelling results show that teaching engagement explains almost all the difference in students' attitudes towards mathematics at the class level. The multilevel structural equation modelling further identifies the indirect effects of SBL on ATM via its influence on student-perceived EGM at both student and class levels. We can thus accept both the hypotheses we have made at the beginning of this chapter, acknowledging that, in master mathematics teachers' classes, teaching significantly affects students' attitudes towards mathematics and that students' sense of school belonging significantly affects their perception of teaching engagement which in turn affects their attitudes towards mathematics.

Affective outcomes at various education levels have drawn increasing attention from research, with studies starting to weigh up the effects of schooling on teaching which in turn affects affective outcomes (Scherer & Nilsen, 2016). Yet, the mediating mechanism in the schooling-teaching-learning loop is less explored using student-rated schooling and teaching data and in countries (e.g., China) where fewer studies like this have been conducted. Research also rarely scrutinises such mechanisms using data collected from master teachers' classrooms on a large scale. More research is needed to tap into the full range of teaching and learning variance in the global setting (Reynolds et al., 2002) and test mechanisms contributing to non-academic outcomes in mathematics, such as ATM, in addition to the recent findings about the mechanism influencing affective outcomes. The empirical analyses in this chapter contributes to research in this area with the latest evidence, by measuring and linking schooling, teaching and learning outcomes in the affective dimension with an understudied cohort of participants, in a less explored context.

In the following chapter, we move the research lens onto one of the increasingly popular competencies, *metacognition*. More specifically, we situate the competence in the mathematical learning context, measure the metacognitive learning outcomes in mathematics and scrutinise the mechanisms that explain excellence in such outcomes.

6

METACOGNITIVE OUTCOMES

Patterns, variation and mechanisms

Chapter overview

- Measuring and modelling metacognitive learning outcomes in maths
- The conscious and self-regulated learners in maths
- The patterns of metacognitive outcomes in maths
- The variation of metacognitive learning outcomes in maths
- Teaching effects on metacognitive learning outcomes in maths
- The paths towards strong metacognitive performance in maths
- Patterns, variation and mechanisms of metacognitive learning outcomes in maths

> To make an individual metacognitively aware is to ensure that the individual has learned how to learn.
>
> – Garner

The second part of the mathematics survey considers metacognitive learning outcomes in mathematics. In this chapter, we look at the results of the measurement and the teaching-learning models that explain metacognitive learning outcomes. The measurement involves the adaption of an existing general instrument, Jr MAI (Sperling et al., 2002), to the mathematics learning context. The analyses of teaching-learning mechanisms are realised through multilevel modelling and multilevel structural equation modelling.

Metacognition is defined as 'awareness and understanding of one's own thought processes' by the Oxford Dictionary. More broadly speaking, it is thinking about

DOI: 10.4324/9781003127925-6

thinking. In the seminal work of Flavell (1976, p. 232) who coined the research term, metacognition stands for the knowledge of 'one's own cognitive processes and products or anything related to them' and 'the active monitoring and consequent regulation and orchestration of these processes'. A simpler layout of the term is knowledge of cognition and regulation of cognition (Sperling et al., 2002; Veenman et al., 2006).

Findings from the neuroscience indicated a close connection of metacognition with mathematical reasoning and problem solving. The prefrontal cortex (PFC) of the brain is the pivot that connects different parts of the brain, knowledge and strategies, playing a crucial role in task and cognition evaluation as well as higher-order thinking in mathematics and various contexts (Anderson et al., 2011; Bunge et al., 2004; Clark & Dumas, 2016; Fleming et al., 2012; Garrison, 2014). In particular, metacognitive competence is found correlated with the volume of grey matter in the anterior PFC (Teffer & Semendeferi, 2012).

In their 'Education 2030' document the Organisation for Economic Co-operation and Development (OECD) stressed the importance of metacognitive skills in relation to students' learning (OECD, 2018, p. 7). This importance was not surprising, given other sources. For example, in the United Kingdom, the EEF Teaching and Learning Toolkit 'metacognition and self-regulation' is reported as a very effective strategy with substantial evidence strength. A review of the metacognition and self-regulated learning literature by the same organisation confirmed this as well, but noted that metacognition is very much related to actual domain knowledge (Muijs & Bokhove, 2020). In the combination of several definitions of metacognition (Lester et al., 1989; Schoenfeld, 1992) emerge five components: the knowledge base, problem-solving strategies, monitoring and control, beliefs and affects, and instructional practices (Dossey, 2017). Metacognitive skills are regarded as a general person-related property throughout a learning environment, rather than a mathematics-specific phenomenon (Veenman et al., 1997). Yet, the domain-specificity of metacognition also is apparent in the way 'expert problem solvers were able to assess and update their mental representations in familial domains, yet were not any more capable than novices to apply these metacognitive skills in unfamiliar domains' (Desoete et al., 2019, p. 565). In the learning of mathematics, metacognition plays a key role in enabling problem solvers to move beyond the obstacles between what they knew and what they would like to know before, during and after solving a problem (Dossey, 2017). It is thus found to be a strong predictor of mathematical performance (e.g., Desoete, Baten et al., 2019; Desoete, Roeyers et al., 2001; Kuzle, 2018; Ohtani & Hisasaka, 2018; Schneider & Artelt, 2010). However, this relation is not just direct, but also runs indirectly via individual factors, like general intelligence, motivation to learn and 'opportunities' that expose children to learning content (Byrnes & Miller-Cotto, 2016; Byrnes & Miller, 2007; Wang et al., 2013). Schneider and Artelt (2010) reviewed the literature on this topic, concluding that there was a substantial impact of metacognition on mathematics performance.

In addition to the project purposes, we are also interested in seeking answers for some of the existing debates on metacognition and its role in mathematics teaching and learning. Utilising confirmatory factor analysis, multilevel modelling and (multilevel) structural equation modelling, we aim to test the following hypotheses with the metacognitive data:

H_{MC1}: Metacognition is a multidimensional latent construct which umbrellaed knowledge of cognition and regulation of cognition which in turn predict the observed variables/items that they contain;

H_{MC2}: Metacognition can be nurtured by teaching that engages students in learning;

H_{MC3}: Metacognitively oriented teaching influences metacognitive performance both directly (H_{MC3a}) and indirectly (H_{MC3b}) through teaching engagement perceived by students.

In what follows, we first explain the overall design and the methods for measuring and validating the constructs and procedures for testing the hypothesised models. Then, we present the results of the measurement, modelling and model fit where necessary. Findings and indications are discussed upon concluding the chapter.

6.1 Measuring and modelling metacognitive learning outcomes in maths

We adapt the 18-item Jr. MAI questionnaire (version B) (Sperling et al., 2002) by setting each item in the context of mathematics learning. For example, the first item originally says: 'I know when I understand something'; in the adapted version, it reads: 'I know when I understand something in mathematics'.

To test the instrument reliability, we run three explorative factor analyses with the entire metacognitive data (18 items), the knowledge of cognition data (KC, nine items) and the regulation of cognition data (RC, nine items). The results indicate that the instrument as a whole has a good internal reliability ($\alpha = .907$), and so do its two dimensions, knowledge of cognition ($\alpha = 0.841$) and regulation of cognition ($\alpha = 0.844$). To test the two-dimensional model of the metacognition construct (H_{MC1}), we run a second-order confirmatory factor analysis, with KC and RC umbrellaed by MC and the nine KC or RC items predicted by KC or RC respectively.

The metacognitive learning outcomes are discussed in terms of patterns and variation. General patterns are explored with descriptive statistics and pairwise correlations. Metacognitive variation between gender and grades is observed, using t-tests or ANOVA. Considering the differences between the number of lower graders (Grades 2–4) and that of upper graders (Grades 5–6), we use the bootstrapping method with 1,000 samples in ANOVA.

To unveil the mechanisms underpinning metacognitive performance in mathematics, we explore the hypothesised teaching-learning mechanisms underpinning metacognitive learning outcomes through multilevel modelling and multilevel structural equation modelling. First, to estimate the effects of teaching on meta-cognitive performance (H_{MC2}), we ran a series of two-level models on metacognition, with age, gender and SES controlled for in the full model. Then, we ran two MSEM (211) models to check the effects of ISTOF5 on metacognition as mediated by student-perceived teaching engagement (H_{MC3}), with the last/second model controlling for student age, gender and SES. The measurement parts of the structural models are a second-order CFA of the MC data, a CFA of the EGM data at level 1 and a CFA of the ISTOF5 at the level 2.

6.2 The conscious and self-regulated learners in mathematics

Through the analyses of the metacognitive data, we are able to understand how well children are able to know and regulate their cognition in the processes of mathematics learning (Table 6.1).

Metacognitively strong learners in mathematics know their cognition/thinking better. Learners who understand their cognition are able to know it if they really understand the content (MC01). They can also understand others' thinking or intention easily, for example, the expectation of their maths teacher (MC04). They know the kind of things that would lead to better learning, such as the strategy that works before (MC03) and prerequisite knowledge (MC05). To reach optimal mathematics learning, they know that they need to make proper use of their own strengths (MC13) and get themselves interested in the subject (MC12). A better

TABLE 6.1 Descriptive statistics of metacognitive learning outcomes in mathematics

Item	Sequence	N	Min	Max	Mean	SD
KC1	MC01	3025	1	5	4.28	0.79
KC2	MC02	3025	1	5	4.13	0.88
KC3	MC03	3020	1	5	4.04	0.95
KC4	MC04	3016	1	5	4.08	1.01
KC5	MC05	3016	1	5	3.70	1.00
KC6	MC14	3020	1	5	3.90	1.03
KC7	MC12	3013	1	5	4.50	0.81
KC8	MC13	3015	1	5	3.85	1.01
KC9	MC16	3018	1	5	3.45	1.11
RC1	MC06	3020	1	5	3.73	1.06
RC2	MC07	3023	1	5	3.64	1.15
RC3	MC08	3019	1	5	3.88	1.09
RC4	MC09	3014	1	5	3.86	1.10
RC5	MC10	3018	1	5	3.95	1.04
RC6	MC11	3014	1	5	4.39	0.83
RC7	MC15	3015	1	5	4.24	0.96
RC8	MC17	3016	1	5	3.54	1.20
RC9	MC18	3022	1	5	4.15	1.00

understanding of one's cognition leads to efficient action taking (MC02) and automatic strategy making (MC16).

Strong metacognitive performance means that learners have better regulation of cognition in the process of mathematics learning. Because of the better regulation, children are good at comparing different methods during problem solving and making sense of the mathematics to hand so as to select a best solution (MC06; MC08). They reflect upon their thinking, strategies and the demand or quality of work before (MC09; MC18), during (MC15; MC10) and after (MC07; MC17) their mathematics learning. They make evaluations about their strategies and build their learning on it (MC17).

6.3 The patterns of metacognitive learning outcomes in maths

As can be found in Table 6.1, on a scale of 5, the children from master maths teachers' classes have an average score of 4.0 (SD = 0.6) in metacognition, with their knowledge of cognition being slightly higher (M = 4.1, SD = 0.6) than the regulation of cognition (M = 3.9, SD = 0.7). In half of the metacognitive items, student scores range from 3.95 to 4.5, with the SD ranging from 0.79 to 1.04; in the other half, the scores range from 3.45 to 3.90, with the SD ranging from 1.00 to 1.20. The skewness and kurtosis values fall within the range of (-1, 1), suggesting a normal distribution. The item mean scores are quite high, in comparison with findings from former studies where the Jr MAI was developed and/or utilised (Ning, 2016; Sperling et al., 2002).

6.4 The variation of metacognitive learning outcomes in mathematics

To observe the variation of metacognitive learning outcomes in mathematics, we consider the effects of age (grade), gender and SES.

In line with existing literature, on the metacognitive scale of 5, children in lower grades (M = 3.82, SD = 0.66) lag behind their senior peers (M = 3.97, SD = 0.62) by 0.15. This means a modest effect posed by the schooling stage: $t(3024) = 4.28, p < 0.001$, Cohen's $d = 0.23$. The bootstrapping method with 1,000 samples in ANOVA further indicates that Grade 2 has significantly higher scores than Grades 5 and 6 in overall metacognition $\left(F(4,3023) = 6.16, p < 0.001, \eta_p^2 = 0.01 \right)$ and that each higher grade outperforms any lower grade in the knowledge of cognition $\left(F(4,3023) = 36.24, p < 0.001, \eta_p^2 = 0.02 \right)$. Nevertheless, the regulation of cognition sees no significant difference between any two grades. The post hoc tests indicate that knowledge of cognition yields absolute difference between any pair of lower grades (2–4) and upper grades (5–6). The effect sizes (i.e., η_p^2) identified in the case of significant differences were, however, almost negligible. The small effect of grades on metacognition coincides with the significant and relatively small correlation $\left(r = 0.10, p < 0.01 \right)$ between age and metacognitive scores. The pairwise correlation also suggests a differential effect of age on knowledge of cognition $(r = 0.13, p < 0.01)$ and regulation of cognition $(r = 0.06, p < 0.01)$.

With a mean of 3.99 $(SD = 0.62)$, girls appeared to be metacognitively stronger than boys $(M = 3.91, SD = 0.65)$, and the t-test showed a significant but weak effect of gender: $t(3022) = 3.42, p < 0.001, d = 0.13$. The gender difference found in the current study echoes similar findings in some of the studies (Daniel et al., 2016; King & McInerney, 2016) but not all (e.g., see Metallidou & Vlachou, 2007).

Socioeconomic status is generally found related to metacognitive performance in studies based on data collected in the nations other than China (Akyol et al., 2010; Daniel et al., 2016; Pappas et al., 2003). The current study replicates this finding in the Chinese context where student SES is indeed positively correlated with their performance in metacognition $(r = 0.19, p < 0.01)$, knowledge of cognition $(r = 0.23, p < 0.01)$ and regulation of cognition $(r = 0.13, p < 0.01)$.

Combining the findings, it seems to suggest that the difference of metacognitive performance between various groups, if any, tends to exist in the dimension of knowledge of cognition. The difference in regulation of cognition is either smaller or insignificant.

6.5 Teaching effects on metacognitive learning outcomes in maths

We have hypothesised at the beginning of this chapter that teaching contributes to metacognitive outcomes (H_{MC2}). Now with metacognitive outcomes measured and teaching and background variables readily available, we can now test this hypothesis using multilevel modelling.

To partition the variance of metacognitive outcomes between the student and teacher levels, we run three multilevel models. The first model (the null model, MC_{MLM0}) tests the proportion of variance explained at the teacher level. With an ICC of 0.12, the model justifies the rationale for multilevel modelling, by suggesting that 12% of the variance in student metacognitive learning outcomes in mathematics is situated at the teacher/class level. We are thus confident to continue with the rest of the modelling.

The second model (MC_{MLM1}) estimates the effects of the three background variables – student age, gender and SES – on their metacognitive performance. We can see in Table 6.2 that the variance in metacognitive learning outcomes drops by 5.7% at the student level and 20% at the teacher/class level.

There is much variance left unexplained in the model MC_{MLM1}, so we carry on modelling the data by adding the teacher level predictor, the aggregated EGM (L2EGM), in the full model (MC_{MLM2}) which is expressed in the following equations:

Level-1 equation:

$$MC_{ij} = \beta_{0j} + \beta_{1j} Age_{ij} + \beta_{2j} Gender_{ij} + \beta_{3j} SES_{ij} + e_{ij},$$

TABLE 6.2 The two-level model of metacognitive learning outcomes in mathematics

Models	MC_{MLM0}		MC_{MLM1}		MC_{MLM2}	
Fixed part (Intercept)	Coefficients 3.95***	ES	Coefficients 2.67***	ES	Coefficients -0.18	ES
Student level						
Age			0.05**	0.22	0.05**	0.22
Gender			0.06**	0.27	0.06**	0.27
SES			0.02***	0.09	0.02***	0.09
Teacher level						
L2EGM					1.1***	4.93
Random part	Variance	SD	Variance	SD	Variance	SD
Student level (σ^2)	0.35	0.595	0.33	0.576	0.33	0.576
Teacher level (τ_{00})	0.05	0.223	0.04	0.208	0.02	0.131
ICC	0.12		0.11		0.05	
$N_{Teacher}$	67		67		67	
$N_{Observation}$	3028		2872		2872	
Deviance	5576.8		5104.8		5057.2	
Marg R^2/Cond R^2	0.000/0.123		0.057/0.165		0.115/0.158	
Variance Reduction from Null (%)						
Student level			5.7%		5.7%	
Teacher level			20.0%		60.0%	

Note: $ES = \dfrac{coefficient}{SD_{\tau_{00}NullModel}}$. *EGM* = student-perceived teaching engagement in mathematics class measured with the scale ASBM02A-J in TIMSS 2015 Student Questionnaire for Grade 4 Mathematics (IEA, 2014). ** $p < 0.01$. *** $p < 0.001$.

Level-2 equations:

$$\beta_{0j} = \gamma_{00} + \gamma_{01} L2EGM_j + u_{0j},$$
$$\beta_{1j} = \gamma_{10},$$
$$\beta_{2j} = \gamma_{20},$$
$$\beta_{3j} = \gamma_{30}.$$

Combined equation:

$$MC_{ij} = \gamma_{00} + \gamma_{10} Age_{ij} + \gamma_{20} Gender_{ij} + \gamma_{30} SES_{ij} + \gamma_{01} L2EGM_j + u_{0j} + e_{ij}$$

In the above equations, MC_{ij}, Age_{ij}, $Gender_{ij}$ and SES_{ij} are the metacognition score, age, gender and SES of student i in class j respectively; $L2EGM_j$ is the level-2 predictor, the aggregated EGM, in class j; e_{ij} and u_{0j} are the residual error terms at levels 1 and 2 respectively.

As shown in Table 6.2, the full model (MC_{MLM2}) reduces the teacher-level variance by 60%, with the reduction of student-level variance (5.7%) being the same as in the former model (MC_{MLM1}). The class-level perception of teaching engagement manifests a significant coefficient of 1.1 on student metacognitive scores, with a much larger effect size (4.93) than *Gender* (0.27), *Age* (0.22), and *SES* (0.09).

Between the two conditional models (i.e., MC_{MLM1} and MC_{MLM2}), the influence of age, gender and SES remains stable, but the level-2 variance dropped by 40% in the latter model where the teaching variable is included. In line with literature and echoing the pairwise correlation and *t*-test results in section 6.4 of the chapter, age plays a significantly positive role in metacognitive performance ($\gamma_{10} = 0.05$, $p < 0.01$); girls scored relatively higher than boys ($\gamma_{20} = 0.06$, $p < 0.01$); the effect of SES is almost negligible albeit significant ($\gamma_{30} = 0.02$, $p < 0.001$). Whilst age, gender and SES all manifest significant effects on metacognition, all the effect sizes (0.22, 0.27, and 0.09) are quite small, in comparison with the size of teaching effect (4.93). The model suggests that, at the teacher/class level, if student-perceived teaching engagement at the class level increases by one point on a scale of 3, students' metacognitive performance moves 1.1 point up on a scale of 5 ($\gamma_{01} = 1.1$, $p < 0.001$). Teaching does make a significantly strong impact on metacognitive learning outcomes in mathematics (H_{MC2}).

To understand the paths from observed teaching to metacognitive learning outcomes via perceived teaching, we now move on to model the metacognitive mechanisms using multilevel structural equation modelling.

6.6 The paths towards strong metacognitive performance in maths

Two models, MC_{MSEM0} and MC_{MSEM1}, were run on the relation between metacognitively oriented teaching (ISTOF5) and metacognitive performance (MC) as well as the mediating role of perceived teaching engagement (EGM) in this relation (H_{MC3}). Whilst both models put such correlation and mediation at the centre of their attention, the former excludes and the latter includes the background variables (student age, gender and SES).

As the fit indices show in Table 6.3, the three measurement models all manifest an acceptable-to-good fit to the data. The ISTOF5 has a satisfying χ^2 test and χ^2 / df value, because the sample size for the teaching data is much smaller than the size of the student sample which is likely to have led to the significant χ^2 test for the CFA models of the MC and EGM. Judging by the MSEM indices, both models had an acceptable CFI and TLI as well as a good RMSEA and SRMR. With smaller values of the χ^2 / df, AIC and BIC, the MC_{MSEM1} demonstrates a better fit to the data.

We now zoom in to see the effect sizes of various paths between the variables. At the student level of the unrestricted model (MC_{MSEM0} in Table 6.4), teaching engagement significantly predicts metacognitive performance ($\beta = 0.58$, $p < 0.001$), with the effect dropping a little ($\beta = 0.53$, $p < 0.001$) in the conditional model (MC_{MSEM1}). At the teacher level, observed metacognitive teaching (ISTOF5) poses

TABLE 6.3 Model fit indices related to the models on metacognition

MSEM: IF5→EGM→MC

Model	χ^2	df	p	χ^2/df	CFI	TLI	RMSEA
CFA							
IF5	1.557	2	0.459	0.779	1.000	1.012	0.000
MC	1132.134	135	0.000	8.386	0.943	0.936	0.051
EGM	820.425	35	0.000	23.441	0.924	0.902	0.089
MSEM							
MC_{MSEM0}	3121.127	724	0.000	4.311	0.909	0.901	0.035
MC_{MSEM1}	3207.422	802	0.000	3.999	0.904	0.896	0.034

Model	$SRMR_w$	$SRMR_b$	AIC	BIC	N	Note
CFA						
IF5	0.022				67	4 indicators
MC	0.033				2895	2nd-order CFA: MC=~KC+RC
EGM	0.044				2829	10 items
MSEM						
MC_{MSEM0}	0.042	0.127	159673.292	160523.090	2701	2-level without controls
MC_{MSEM1}	0.044	0.123	151481.123	152359.397	2579	2-level + age + gender + SES

Note: MC_{MSEM0} is the model without controlling for background variables. MC_{MSEM1} is the model controlling for age, gender and SES. CFI = comparative fit index. TLI = Tucker-Lewis index. RMSEA = root mean square of approximation. SRMR = standardised root mean square residual.

TABLE 6.4 Path coefficients for the multilevel structural equation models of metacognition

Model	Student level	Teacher/Class level		
	MC~EGM	MC~EGM	MC~ISTOF5	EGM~ISTOF5
MC_{MSEM0}	0.58***	0.60***	0.10	0.55***
MC_{MSEM1}	0.53***	0.83**	-0.11	0.34*

Note: * $p < 0.05$. ** $p < 0.01$. *** $p < 0.001$.

a positive effect on teaching engagement (MC_{MSEM0}: $\beta = 0.55, p < 0.001$; MC_{MSEM1}: $\beta = 0.34, p < 0.05$) which in turn affects metacognitive performance (MC_{MSEM0}: $\beta = 0.60, p < 0.001$; MC_{MSEM1}: $\beta = 0.83, p < 0.01$). At the class level, ISTOF5 predicts metacognition performance indirectly via its effect on teaching engagement, with the direct effect being insignificant.

In summary, whilst the MSEM results do not support the hypothesis that metacognition-focused teaching influences metacognitive performance directly (H_{MC3a}), we can accept the hypothesis H_{MC3b} that metacognitive performance is indirectly affected by metacognitively oriented teaching via the mediation of student-perceived teaching engagement.

6.7 Patterns, variation and mechanisms of metacognitive learning outcomes in maths

In this chapter, we have tested three hypotheses, in addition to descriptive statistics of metacognitive outcomes in mathematics.

Metacognitively strong learners have a better understanding of their own status of cognition and a better regulation of their thinking, planning and learning (Flavell, 1976, 1979; Sperling et al., 2002). Our findings in this chapter indicate that in general the master maths teachers' students demonstrate quite good performance in metacognition (MC), having good knowledge (KC) and regulation (RC) of their own cognition during the process of mathematics learning. On a scale of 5, they have an average score of 3.96 ($SD = 0.63$) on MC, with a slightly higher score on KC (3.99 ± 0.64) than RC ($3.92 \pm 0,70$). The mean scores are much higher than the results from formers studies using the same instrument (Ning, 2016; Sperling et al., 2002). About the variation of the metacognitive outcomes, a significant effect comes from age $(r = 0.10, p < 0.01)$ and SES $(r = 0.19, p < 0.01)$, and girls do better than boys (3.99 vs. 3.91 out of 5, $t(3022) = 3.42, p < 0.001, d = 0.13$).

To check the validity of the Jr MAI's intended structure, we ran a second-order CFA model in which two nine-item factors are umbrellaed by the overarching latent variable – metacognition. The results indicate a good fit of the model to the data. We therefore are confident that metacognition, as measured with the adapted mathematics-oriented version of the Jr MAI (Sperling et al., 2002), can be indeed modelled as a latent variable consisting of two dimensions – knowledge of cognition and regulation of cognition – which in turn can be properly estimated with the nine items affiliated to them. The H_{MC1} is thus accepted.

Our multilevel modelling of the metacognitive performance further shows that teaching engagement explains 40% of variance in metacognitive performance at the class level, controlling for age, gender and SES. The results thus suggest we accept the H_{MC2}. The multilevel structural equation modelling indicates that metacognitively oriented teaching (ISTOF5) predicts student-perceived teaching engagement which in turn predicts metacognitive learning outcomes. The results accept the H_{MC3b} (indirect effect) but reject the H_{MC3a} (direct effect).

Research has rarely given exclusive attention to metacognitive performance of children taught by expert mathematics teachers. The mechanism behind metacognition is rarely explored with classroom data using both the MLM and MSEM approaches. This is even rarer in the context of mathematics teaching and learning. The chapter contributes to the field with respect to, though not limited to, these gaps.

A key question in the field is the relationship between metacognition and cognition. According to Pressley and Harris (2006), it is very hard to have knowledge about how competent one is or know how best to learn in a domain without domain-specific knowledge. For example, students need to know what key concepts are in a subject area, and how they relate to one another. This also includes a judgement of task difficulty, which may also interact with the affective components (e.g., self-efficacy, self-concept and confidence). Likewise, it is unlikely a student would know what (metacognitive) skills to use to solve a problem without having a (cognitive) method to do so. It is here where we can see that knowledge of cognition and regulation of cognition go hand-in-hand, something which is expressed in the way we measure metacognition in the Jr MAI. An important reason to not solely focus on knowledge of metacognition, is that metacognitive knowledge also can be wrong (we can underestimate the time we need to memorise something, for example), with students' metacognition being suboptimal in terms of effectiveness and efficiency. It is here where schooling plays an important role, as schooling can improve both metacognitive knowledge and skills through teaching and learning (Veenman et al., 2006).

In the context of mathematics education, metacognition should not be seen as a stable characteristic of an individual learner, but as something that interacts with the task at hand. It is context-specific which is why we adapted our instruments to the specific context of mathematics learning. Metacognition has been found to be quite context-dependent, which means that a student who shows strong metacognitive competence in one task or domain may be weak in another, and metacognitive strategies may be differentially effective depending on the specific task, subject or problem tackled (Kim et al., 2013). In the current project, the mathematically adapted Jr MAI shows a good fit of second-order model to the data, suggesting the possible, albeit not exactly affirmative, fact that metacognition can be subject-specific.

In previous research, strong metacognition has been found correlated with better cognitive performance in mathematics (Desoete et al., 2019; see some examples in Muijs & Bokhove, 2020; Zhao et al., 2019). We will explore the potential connection between teaching, metacognition and cognition in mathematics in the next chapter.

7

ACHIEVEMENT IN MATHEMATICS

Patterns, variation and mechanisms

Chapter overview

- Measuring and modelling cognitive learning outcomes in mathematics
- Patterns and variation of mathematics performance
- Children better prepared for algebra and advanced mathematics
- Flexibility and proficiency in proportional reasoning
- Teaching effects on achievement in mathematics
- Paths towards strong mathematics performance
- Patters, variation and mechanisms of cognitive learning outcomes in maths

> I don't have any magical ability. I look at a problem, play with it, work out a
> strategy.
>
> — Terence Tao

In this chapter, we look at the measure of and the mechanisms behind mathematical achievement using the teaching and learning data collected in Grades 5 and 6. Instead of testing the students with all domains of the primary mathematics, we focus on the most challenging topics in school mathematics – rational numbers and proportional reasoning. This allows us to carry out a more thorough assessment of mathematical competence in this domain.

Amongst all school mathematics topics, rational numbers and ratios are regarded as 'the most protracted in terms of development, the most difficult to teach, the most mathematically complex, the most cognitively challenging, the most essential

DOI: 10.4324/9781003127925-7

to success in higher mathematics and science, and one of the most compelling research sites' (Lamon, 2007, p. 629). These topics were amongst the key topics systematically studied by the CSMS project (Hart et al., 1981) and by the recent ICCAMS project (Hodgen et al., 2010) utilising test items developed by the former in the 1970s. In the CSMS project, only 12% of students in secondary Year 1 and 20% in Year 4 made it explicit that there were an infinite number of numbers between two seemingly adjacent numbers, 0.41 and 0.42 (Hart et al., 1985b).

Rational numbers are defined as 'elements of an infinite quotient field consisting of infinite equivalence classes', whilst fractions are 'the elements of the equivalence classes' (Behr et al., 1992, p. 296). A rational number can be written in the form of a fraction (or ratio or two integers), such as $\frac{a}{b}$, where a and b are both integers and b is nonzero (Houston, 2009). Whole numbers can be written as fractions as well, but in the MasterMT project, we focus primarily on those written in the forms of fractions. Decimals in the current study refer narrowly to those that are elements of the set of Rational Numbers, excluding both non-terminating repeating decimals and non-terminating non-repeating decimals. The former may be represented as a fraction (e.g., $0.\dot{3} = \frac{1}{3}$), whereas the latter is irrational (e.g., π, e, and $\sqrt{2}$). Within this boundary, decimals are a special type of fractions that have 10^n as denominators, for example, $0.2 = \frac{2}{10}$, $0.02 = \frac{2}{10^2}$, $0.002 = \frac{2}{10^3}$. Here, we use 'decimals and fractions' and 'rational numbers' interchangeably. Proportional reasoning is the cognitive status 'where student presents valid reasons in support of claims made about the structural relationships that exist when two ratios are equivalent' (Lamon, 1995, pp. 172–173).

The more overt connections between rational numbers and algebra may be observed in a simple equation like $y = ax$ where the slope is the nonzero constant, a. This slope value is equal to the magnitude of the fraction, y / x, where $x \neq 0$, as well as the value of the ratio, $y : x$, where $x \neq 0$. The equation holds for any pair of y and x $(x \neq 0)$ as long as the fractions/ratios formed by these two variables are elements of the set of equivalent fractions/ratios with the same magnitude/value, a (Peck, 2020). In this sense, the facts that fraction and ratio can show both the relation between two quantities and a value (quotient) are the two sides of one coin that connects rational numbers and simple equations like $y = ax (a \neq 0)$. It is fractions representing relations between quantities and decimals as unidimensional magnitudes that were found to have significantly predicted algebra performance (DeWolf et al., 2015).

The National Mathematics Advisory Panel (NMAP, 2008) found fractions to be one of the crucial areas in preparing students for algebra learning and 'the most important' yet underdeveloped area amongst American students, concluding that 'the teaching of fractions must be acknowledged as critically important and improved before an increase in student achievement in algebra can be expected'

(NMAP, 2008, p. 18). Indeed, the crucial role of rational numbers and ratios is evident. Learning outcomes on the topics have been found posing a lasting effect on later mathematics performance in general and algebra in particular, even when knowledge in other maths topics, general IQ and SES were controlled for (Siegler et al., 2012). Such lasting effects of rational number fluency on algebra competence were even found amongst undergraduates, long after their completion of learning the content (Hurst & Cordes, 2018). Unsurprisingly, these topics together have been one of several focal areas in international surveys aiming at fundamental and/ or higher-order maths competencies, such as TIMSS (Mullis et al., 2016, Exhibit 2.1) and PISA (OECD, 2019, p. 105, Table I.6.1).

In the MasterMT project, the mathematics test is thus focused on the most challenging and important topics in school mathematics – rational numbers and proportional reasoning. In addition to the test results seen through macro, meso- and micro lenses (Miao et al., 2022), this chapter presents our findings on and about mathematics tests in two ways. First, we look at patterns and variation of student cognitive outcomes in mathematics. Second, we look closely at the teaching-learning (TL) mechanisms through the lenses of multilevel modelling and multilevel structural equation modelling. In addition to the utilisation of descriptive statistics and Rasch modelling, we delve into the test and the TL mechanisms, testing the following hypotheses informed by existing research:

H_{M1}: Student age (H_{M1a}), gender (H_{M1b}) and SES (H_{M1c}) significantly affects their cognitive outcomes in mathematics;

H_{M2}: Teaching significantly affects cognitive outcomes in mathematics in spite of the effects of age, gender and SES;

H_{M3}: Metacognitive oriented teaching affects cognitive outcomes in mathematics both directly and indirectly through its effect on metacognitive outcomes in mathematics.

7.1 Measuring and modelling cognitive learning outcomes in mathematics

Whilst the detailed methods of measurement can be found in Chapter 3, here we have a quick recap on how we analyse the maths test data and model the hypothesised mechanisms.

As explained in Chapter 3, the cognitive outcomes in mathematics were measured with three test booklets (G5A, G5B and G6) consisting of a total of 28 items adapted from the CSMS project. Each student in Grades 5 and 6 completed one of the three tests. Tests were analysed with the Rasch model to estimate person ability and item difficulty; the equating of the three tests' data was performed with the seven common items working as anchor items. The equated test scores were linearly transformed into scores with a mean of 500 and *SD* of 100.

Our initial analyses of the test data focus on the patterns of AIM on the item, topic and test levels and the variation of student performance by age, gender and SES. After these, we move on to systematic analyses of the relationship between teaching and learning through multilevel modelling and multilevel structural equation modelling.

7.2 Patterns and variation of mathematics performance

Amongst the 2,642 fifth and sixth graders who have an average score of 500 in mathematics, the higher graders (541 ± 109) perform significantly better than the lower graders (458 ± 68) by an average difference of 83 points $(t(2222) = 23.53, p < 0.001, \text{Cohen's } d = 0.94)$. As we hypothesised, age $(r = 0.33, p < 0.01)$ is positively correlated with cognitive outcomes in mathematics (H_{M1a}). SES has a similar correlation with maths $(r = 0.23, p < 0.01)$, as we anticipated (H_{M1c}).

As often found in various assessments, including those from international large-scale studies, there is a gender gap amongst fifth graders, with girls slightly lagging behind boys $(MD = 467 - 449 = 18, t(1306) = 4.58, P < 0.001, \text{Cohen's } d = 0.27)$. The encouraging finding is that the significance of the gap vanishes in Grade 6 $(MD = 545 - 539 = 6, t(1320) = 1.03, p = 0.30)$. Thus, the H_{M1} proves to be partially correct. Our anticipation is the extra-year quality teaching might contribute to the gap closure.

Across the two grades, an average of 78% of students succeeded in decimals and fractions. The Rasch analyses show that decimals are slightly easier than fractions and that ratios and proportions are much harder than decimals and fractions (Miao et al., 2022). Across grades, the average success rate is 78% on decimals and fractions and 52% on ratio. In comparison with the secondary school students in the CSMS study (Hart et al., 1985a; Hart, Brown et al., 1981), the primary school students in the current study had achieved a much higher correct rate on each of the 28 items (Figure 7.1). The current gap might be wider, given the considerable decline of English students' performance on these topics found by the ICCAMS project 30 years later (Hodgen et al., 2010). Nevertheless, it should be acknowledged that there are sample differences between the two English studies and the MasterMT. The English results were generated from nationally representative samples of English secondary students, whereas the sample for the current study came from classes of master mathematics teachers in primary schools of urban China.

In terms of level proficiency (see Figure 3.3), there is a higher proportion of students reaching the top levels in decimals, fractions and ratio and proportions than that of the CSMS cohort (Figure 7.2). Research and practice suggest children often struggle with tasks such as fraction division, reasoning proportionally beyond doubling and tripling, and understanding the existence of an infinite number of numbers between two seemingly adjacent fractions/decimals, for example, between 1/3

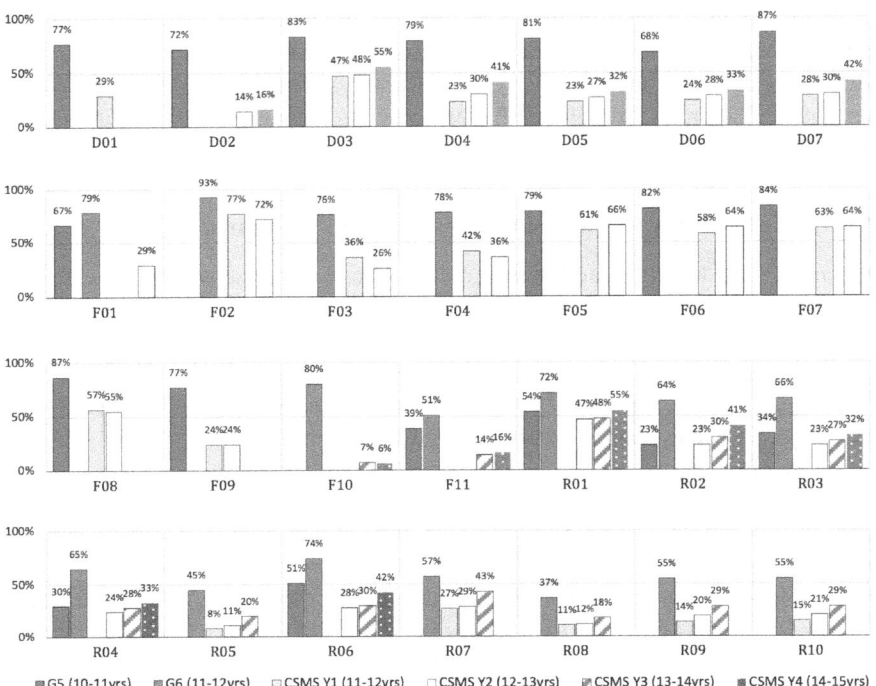

FIGURE 7.1 Item facilities on the CSMS benchmark

Note: G5 = Grade 5 (aged 10–11); G6 = Grade 6 (aged 11–12). CSMS = the Concepts in Secondary Mathematics and Science project (Hart et al., 1981). Y1 = Secondary Year 2 (aged 11–12) in England in the 1970s. Y2 = Secondary Year 2 (aged 12–13). Y3 = Secondary Year 3 (aged 13–12). Y4 = Secondary Year 4 (aged 14–15).

and 2/3, or 0.11 and 0.12. These do not seem as difficult for the students from the master teachers' classes. Rational numbers and proportional reasoning have been found as predictors of algebraic performance (Hurst & Cordes, 2018; Siegler et al., 2012). Having not attempted to replicate this finding, we could, however, see the early sign of transition from proportional thinking to algebraic thinking amongst sixth graders who tackle ratio tasks using algebraic methods, in addition to arithmetic.

The fifth and sixth graders from the master maths teachers' classes in China show a higher degree of readiness for learning algebra and more advanced mathematics at a younger age than their senior peers in the CSMS and ICCAMS studies where the same test items were utilised. With flexibility and proficiency in reasoning proportionally, the upper primary children in the current study show a deeper understanding of the big idea behind the phenomena of rational numbers and ratios. Of course, when interpreting the results, we should constantly remind ourselves of

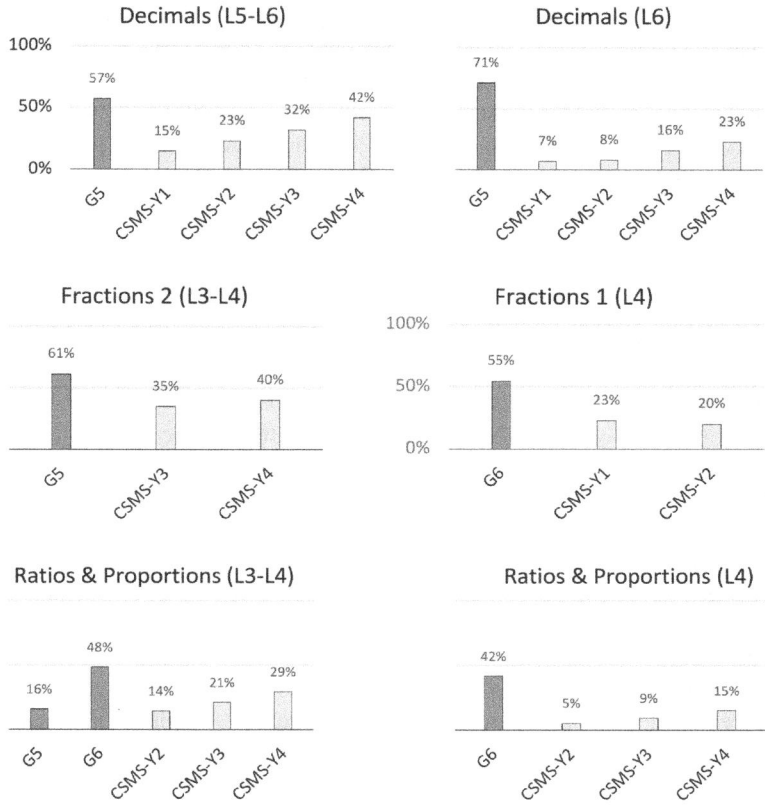

FIGURE 7.2   Top-level rational and proportional reasoning on the CSMS benchmark

the fact that we are comparing primary students taught by master maths teachers in China with the national samples of Key Stage 3 in England in 1979 and 2009.

7.3   Children better prepared for algebra and advanced mathematics

Overall, students in the current study performed relatively well in rational numbers and proportional reasoning – the pivotal topics that affect algebraic learning and overall mathematics in secondary and higher education (Hurst & Cordes, 2018; Siegler et al., 2012). Chronologically and cross-nationally, these Chinese primary children had substantially outperformed their senior peers in English secondary schools back in 1976 (Hart et al., 1985) and 2008 (Hodgen et al., 2010). They appeared to be better prepared for the learning of algebra and more advanced maths. Again, it should be noted that the Chinese sample came from master mathematics teachers' classes where better learning is likely to happen, albeit not definitely easily so at such young ages. The results, however, do suggest that children can reach a higher level of mastery in the most challenging topics of school mathematics at a younger age. It is thus reasonably possible for other children of a similar age to break the performance ceiling that hinders their development. Lifting *learning* upwards offers more developmental opportunities to learners than attempting to develop *learners* through the somewhat fixed developmental stages (Biggs, 1982).

7.4   Flexibility and proficiency in proportional reasoning

In the current study, the 5th and 6th graders from Chinese cities demonstrated a high level of proficiency in the top-level items of the assessed topics and were able to use the connections between division, fractions and ratios to solve problems flexibly. Those successful on the most difficult items were able to not only see fractions as numbers but also as magnitudes dependent on two division-related quantities – the numerator and the denominator. Likewise, in their mind, ratios were similar to fractions which not only represented the relationship between two terms but also carried values – a concept similar to the magnitude of fractions. Moreover, both ratios and fractions share fundamental traits with division which not only results in a value but also denotes the relationship between two quantities (Cai & Wang, 2006; Moseley et al., 2007). Such proficiency shows these children's ability in thinking proportionally and switching smoothly at their will between different representations and procedures according to specific problem situations. To them, division, fractions and ratios are just phenomena (Freudenthal, 1983) of one big idea (Askew, 2013). Choosing the appropriate unit and partitioning is the key in reasoning about rational numbers. Proficient learners can switch easily between various 'phenomena' of rational numbers and tackle both abstract and real-world problems through representing quantities properly and modelling relationships between them accurately with appropriate mathematical models.

The learning of these topics may be greatly enhanced by teaching with a particular emphasis on the between-topic connections after the mastery of each specific topic, such that learners can see the bigger idea behind the interconnected topics gradually emerging (Askew, 2013; Lamon, 1993). This links rational numbers and ratios closely to algebra where building relations between variables is central. Echoing 'relational thinking' in work by Empson et al. (2011, pp. 413–425), such connectedness enriches and strengthens children's learning in fractions and extends their thinking algebraically.

Unsurprisingly, in the Chinese curriculum and amongst Chinese teachers, it is treated as a matter of fact that there exist four sets of triple counterparts weaving together the three concepts, fraction, ratio and division: (1) the antecedent (i.e., the first term) of a ratio is conceptually similar to the dividend in division and the numerator of a fraction; (2) the colon between two terms of a ratio is conceptually similar to the division sign and the fraction line; (3) the consequent (i.e., the second term) of a ratio is conceptually similar to the divisor in division and the denominator of a fraction; (4) the value of a ratio is conceptually similar to the quotient of division and the value or magnitude of a fraction (Cai & Wang, 2006). However, the curriculum in itself does not guarantee good performance amongst all learners, as shown in recent studies on performance of Chinese students in fractions and decimals (Jiang et al., 2021; Liu et al., 2014). Intended curriculum is not implemented curriculum (Valverde et al., 2002). Otherwise, in a more centralised system like China, teachers would be achieving the same excellence and so would students. But that is simply not the case.

Given the connected nature of the assessed topics, it is crucial for children to have a full vision and thorough understanding of all possible representations or *phenomena* of rational numbers and ratios (Freudenthal, 1983). Such a proficiency would lead to children's ultimate mastery of knowledge and skills in these inter-related subdomains and thus lay solid foundation for their proficiency in more advanced mathematics. Teachers' emphases of the connectedness of rational numbers and ratios in class would help nurture higher-order thinking about the higher-order knowledge of primary mathematics (Cai & Wang, 2006; Howe et al., 2015). Mathematics is a subject with a strong trait of connectedness within itself, and the benefit of teacher/teaching effort given to building a robust *net* of knowledge is evident in the empirical literature (Askew et al., 1997; Ma, 1999). It is essential to teach/learn ratios with a wide and rich variety of representations, so that learners are able to get thorough understanding of the properties of ratios and proportions, identify the four terms of a pair of equal ratios embedded in various situations, model the relationship between them and solve problems based upon conceptual understanding and procedural clarity (Howe et al., 2015).

Having reflected upon the mathematics test findings, let us now zoom out to see the grander patterns – the mechanisms behind mathematics performance.

7.5 Teaching effects on achievement in mathematics

At the beginning of the chapter, we hypothesised that teaching significantly affects the overall AIM despite the influence of student background variables – age, gender and SES (H_{M2}). To test the hypothesis, we plan to run five 2-level models.

As always, we first run a null model (AIM_{MLM0}) to check if multilevel modelling is necessary. The null model shows an ICC of 0.29, indicating that 29% of variance in mathematics performance is located at the teacher level. This justifies the existence of a multilevel structure in the data.

We then add the three control variables (*Gender*, *SES* and *Age*) in the model AIM_{MLM1} to evaluate the possible effects of student background. Then, we add the two teaching predictors, MQI and ISTOF, separately or together, and get three models: $AIM_{MLM2a-MQ}$, $AIM_{MLM2b-IF}$ and $AIM_{MLM2c-MQIF}$.

We anticipate that adding the two teaching predictors together in the model ($AIM_{MLM2c\text{-}MQIF}$) will likely end up with an insignificant result, due to the overlaps between the two observation systems. The result of the model $AIM_{MLM2c\text{-}MQIF}$ confirms what we thought. When both the MQI and the ISTOF are added (AIM_{MLM2c}), the coefficients become insignificant, indicating a possible collinearity between the two. This makes sense since the correlation between the two is strong ($r = 0.93, p < 0.01$). We therefore consider the AIM_{MLM2a} and AIM_{MLM2b} as the two full models each independently evaluating the effects of teaching on mathematics achievement. In the Table 7.1, we only report these two acceptable full models ($AIM_{MLM2a\text{-}MQ}$ and $AIM_{MLM2b\text{-}IF}$) which are expressed in the following equations:

Level-1 equation:

$$Maths_{ij} = \beta_{0j} + \beta_{1j}Age_{ij} + \beta_{2j}Gender_{ij} + \beta_{3j}SES_{ij} + e_{ij},$$

Level-2 equations:

$$\beta_{0j} = \gamma_{00} + \gamma_{01}Teach_j + u_{0j}$$
$$\beta_{1j} = \gamma_{10},$$

TABLE 7.1 Two-level models of mathematics performance

Models	AIM_{MLM0}		AIM_{MLM1}		$AIM_{MLM2a\text{-}MQ}$		$AIM_{MLM2b\text{-}IF}$	
Fixed part	Coefficient	ES	Coefficient	ES	Coefficients	ES	Coefficient	ES
(Intercept)	501***		393.07***		196.42**		196.78**	
Student level								
Age			3.90		4.95		5.12	
Gender			-14.80***	-0.18	-14.74***	-0.18	-14.72***	-0.18
SES			2.08***	0.02	2.06***	0.02	2.05***	0.02
Teacher level								
MQI mean					42.38***	0.51		
ISTOF mean							41.85**	0.50
Random part	Variance	SD	Variance	SD	Variance	SD	Variance	SD
Student level (σ^2)	7032.58	53.88	6757.74	82.21	6760.88	82.22	6761.39	82.23
Teacher level (τ_{00})	2903.57	83.86	2414.66	49.14	1912.34	43.73	1942.50	44.07
ICC	0.29		0.26		0.22		0.22	
$N_{Teacher}$	59		59		59		59	
$N_{Student}$	2642		2507		2507		2507	
Deviance	31075.4		29385.5		29373.8		29374.9	
Marg R²/ Cond R²	0.000/0.292		0.030/0.286		0.089/0.290		0.086/0.290	
Variance Reduction from Null (%)								
Student level			3.9%		3.9%		3.9%	
Teacher level			16.8%		34.1%		33.1%	

Note: $ES = \dfrac{coefficient}{SD_{\tau_{00}NullModel}}$. AIM = achievement in maths. ** $p < 0.01$. *** $p < 0.001$.

$$\beta_{2j} = \gamma_{20},$$
$$\beta_{3j} = \gamma_{30}.$$

Combined equation:

$$Maths_{ij} = \gamma_{00} + \gamma_{10}\, Age_{ij} + \gamma_{20}\, Gender_{ij} + \gamma_{30}\, SES_{ij} + \gamma_{01}\, Teach_{j} + u_{0j} + e_{ij}.$$

In the equations, $Maths_{ij}$, Age_{ij}, $Gender_{ij}$ and SES_{ij} are the maths score, age, gender and SES of student i in class j respectively; $Teach_{j}$ is the MQI or ISTOF score of the maths teacher in class j; e_{ij} and u_{0j} are the residual error terms at levels 1 and 2 respectively.

As shown in Table 7.1, controlling for age, gender and SES, the model AIM_{MLM1} sees the reduction of 16.8% of the variance on level 2. For mathematics, age is not a significant predictor $(\gamma_{10} = 3.90, p = 0.39)$. Girls score 14.80 points lower than boys $(\gamma_{20} = -14.80, p < 0.001)$, which is consistent with the t-tests discussed earlier. One point increase on the SES index is likely to raise the mathematics performance by 2.08 points $(\gamma_{30} = 2.08, p < 0.001)$ which is not substantial for overall mathematics performance of participants (500 ± 100).

Adding the MQI in the model AIM_{MLM2a}, we see the variance of mathematics achievement on level 2 is reduced by 34.1%. Similarly, when we add the ISTOF, instead of the MQI, as the level-2 predictor, the model MAT_{MLM2b} indicates a reduction of Level-2 variance by 33.1%. In either model, a substantial proportion of variance at level 2 is explained. In addition to the 16.8% of variance explained by the control variables, each teaching predictor further explains 16.3% to 17.3% of variance on level 2, leaving about two-thirds of the level-2 variance unexplained.

The results suggest that even in master mathematics teachers' classes, teaching still makes a considerable difference in learning. They still manage to further push the ceiling of teaching effects, sailing against the sociodemographic current and reducing the expected time length of learning/schooling.

The results also indicate that a more complicated observation system is needed for further reduction of the variance at the teacher level, echoing the recently noted absence of an observation system that could 'catch' all predicators (Charalambous & Praetorius, 2018). The development of a more comprehensive observation system can be a major direction for future research in methodological terms. Nevertheless, before that and within the current project, more explanation can be found in the characteristics of teaching and the processes of PD that are captured in our *qualitative* analyses of the project's data (Chapters, 2, 4, 8 and 9). These qualitative findings offer more insights into the teaching-learning mechanism not only in terms of what works and how it works but also with respect to potential extra points of observation for future research.

Driven by both the unexplained variance and the hypothesised role of metacognition in cognitive performance in mathematics, we decided to explore both the direct and indirect effects of teaching on mathematics achievement, using multilevel structural equation modelling.

7.6 Paths towards strong mathematics achievement

In this section, we present the processes and results of the MSEM analyses on the direct and indirect connections between mathematics teaching and cognitive learning outcomes in mathematics, tapping on the mediating effect of metacognitive outcomes on the relation between the two. This way we are able to test the H_{M3}. In particular, we focus on the effect of metacognitively oriented teaching as measured with the ISTOF5. Unlike the item parcel utilised in the multilevel models in the above section, we treat the ISTOF5 as a latent variable, consisting of four indicators, in this series of MSEM analyses, to make the best use of the strengths of the measurement parts that SEM can offer. Another measurement part is the metacognitive outcomes as a second-order CFA model we discussed in the previous chapter. Acknowledging the multilevel structure of the data, we intend to test multilevel structural equation models. We test the models, first without and then with control variables (age, gender and SES). This results in two models (AIM_{MSEM0} and AIM_{MSEM1}) being tested and each demonstrating an acceptable-to-good fit to the data (Table 7.2).

TABLE 7.2 Model fit indices related to the models on maths

Model	χ^2	df	χ^2/df	p	CFI	TLI	RMSEA
CFA							
ISTOF5	1.557	2	0.459	0.779	1.000	1.012	0.000
MC	1132.134	135	0.000	8.386	0.943	0.936	0.051
MSEM: ISTOF5→MC→AIM							
AIM_{MSEM0}	1518.30	380	4.00	< 0.001	0.924	0.915	0.034
AIM_{MSEM1}	1778.24	434	4.10	< 0.001	0.946	0.940	0.036

Model	$SRMR_w$	$SRMR_b$	AIC	BIC	N	Note
CFA						
ISTOF5	0.022				67	4 indicators
MC	0.033				2895	2nd-order CFA:
						MC = ~KC+RC
MSEM: ISTOF5→MC→AIM						
AIM_{MSEM0}	0.038	0.089	142756.12	143392.49	2536	2-level unconditional
AIM_{MSEM1}	0.058	0.130	134691.45	135339.45	2406	2-level + age + gender + SES

Note: AIM = achievement in maths. AIM_{MSEM0} is the null model without controls; AIM_{MSEM1} is the model controlling for age, gender and SES. CFI = comparative fit index. TLI = Tucker-Lewis index. RMSEA = root mean square of approximation. SRMR = standardised root mean square residual.

TABLE 7.3 Path coefficients for the multilevel structural equation models of mathematics performance

Model	Student level		Teacher/Class level	
	AIM~MC	MC~ISTOF5	AIM~MC	AIM~ISTOF5
AIM_{MSEM0}	0.29***	0.44**	0.03	0.44**
AIM_{MSEM1}	0.28***	0.43**	0.01	0.41*

Note: *** $p < 0.001$. ** $p < 0.01$. *** $p < 0.05$.

As shown in Table 7.3, in the unconditional MSEM (AIM_{MSEM0}), ISTOF5 significantly affects metacognition $(\beta = 0.44, p < 0.01)$ and mathematics performance $(\beta = 0.44, p < 0.01)$ at the teacher level, with mathematics performance being significantly predicted by metacognition at the student level $(\beta = 0.29, p < 0.001)$. With background variables (age, gender and SES) added, the conditional model (AIM_{MSEM1}) finds almost no change in: (1) the effect of ISTOF5 on metacognition at level 2 $(\beta = 0.43, p < 0.01)$, (2) its effect on mathematics performance at the teacher level $(\beta = 0.41, p < 0.05)$ and (3) the effect of metacognition on mathematics performance at the student level $(\beta = 0.28, p < 0.001)$.

Taking into consideration the results of both models, we can conclude that, despite the influence of age, gender and SES, metacognitively oriented teaching (ISTOF5) poses a positive effect on both mathematics achievement and metacognitive performance at the class level and then, at the student level, metacognitive performance poses a positive effect on mathematics performance (Table 7.3).

7.7 Patterns, variation and mechanisms of cognitive learning outcomes in maths

In this chapter, we have looked at patterns and variation of cognitive learning outcomes in mathematics and the teaching-learning mechanisms that potentially explain what works and how it works in the nurture of strong mathematical performers.

With an average score of 500 and an *SD* of 100 across Grades 5 and 6, we could see that sixth graders significantly outperform fifth graders by 83 points $\left(t(2222) = 23.53, p < 0.001, \text{Cohen's } d = 0.94\right)$. Overall, the 2,642 students have an average success rate of 78% on decimals and fractions and 52% on ratio and proportion. According to the CSMS criteria, 57% to 71% of fifth graders have reached the top two levels of decimals; 61% of fifth graders have reached the top two levels of fractions, with 55% of sixth graders demonstrating the top-level proficiency; 16% of fifth graders and 48% of sixth graders have reached the top two levels in ratio and proportion where 42% of sixth graders have reached the top level. Overall, the students on the current study demonstrate better performance than the English secondary students two to five years senior to them in existing studies where the same items were developed and utilised (Hart et al., 1981; Hodgen et al., 2010).

The three background factors each do play a significant role in children's cognitive learning outcomes in mathematics. The general trend is, as documented in the existing literature, that age $(r = 0.33, p < 0.01)$ and SES $(r = 0.23, p < 0.01)$ are two positive correlates of maths performance. Nevertheless, the gender effect plays differently in Grades 5 and 6. Girls underperform boys in Grade 5 (467 vs. 449, $t(1306) = 64.58, p < 0.001, d = 0.27$), but the senior grade sees the disappearance of the gender gap (545 vs. 539, $t(1320) = 1.03, p = 0.30$). We can now accept the H_{M1} that student age, gender and SES do significantly affect their cognitive outcomes in mathematics.

To understand whether teaching still makes a difference in master teachers' class, we have run five models. The results indicate that the teaching predictors, MQI and ISTOF, each explain further 16.3% to 17.3% of variance in mathematics performance, in addition to the 16.8% of variance explained by student background variables. We can thus accept the H_{M2} in that teaching does make a significant difference in students' mathematics performance, albeit the influence of age, gender and SES.

To further explore the direct and indirect effect of teaching on cognitive learning outcomes in mathematics, we have run two multilevel structural equation models. The results indicate that, at the class level, metacognitively oriented teaching (ISTOF5) significantly affects both cognitive and metacognitive outcomes the latter of which in turn positively predict cognitive outcomes at the student level. We are thus able to accept the H_{M3}.

In Chapters 5–7, we have systematically studied three kinds of learning outcomes amongst students taught by master maths teachers and the mechanisms likely explaining the causal relationships. It is time to listen to the teachers and hear what they think about mathematics teaching and learning in the next chapter.

8

MASTERLY MATHEMATICS TEACHING IN THE MASTER MIND

Chapter overview

- Teacher knowledge, beliefs and self-efficacy
- Teachers as living synthesised textbooks
- Scaffolding the learning process
- A connected open system of knowledge and knowledge about knowledge
- Cultivating thorough understanding amongst students
- From hands-on to heads-on, from manipulation to mathematisation
- Aiming for deeper and higher-order thinking and reasoning
- Teaching towards learning to learn with good habits
- Cultivating positive attitudes towards mathematics
- Ready for evolving into master teachers

> Ideas are the source of all things.
>
> – Plato

This chapter is about ideas. It draws on the data from post-lesson interviews with master teachers and the teacher survey data regarding teaching beliefs and self-efficacy as measured with the TALIS 2013 scales (OECD, 2013). More specifically, it captures the mental pictures of master teachers, with respect to their knowledge and beliefs about mathematics and the teaching of it as well as their knowledge and beliefs about learners and mathematics learning.

DOI: 10.4324/9781003127925-8

This chapter is different from Chapter 2 in that it presents a 'high-resolution' yet 'wide-angle' image of master teachers, albeit both chapters take a similar ethnographic approach to researching on this particular 'tribe' of teachers.

Through grounded theory and constant comparisons (Glaser & Strauss, 1967), we code the audio-recorded teacher interview data. Eight common themes emerge amidst teachers' reflections upon the specific lessons they just delivered and mathematics teaching and learning in a broader sense. The themes include (1) teachers as living synthesised textbooks; (2) scaffolding the learning process; (3) demonstrating a connected open system of knowledge and knowledge about knowledge; (4) cultivating thorough understanding amongst students; (5) facilitating transition from hands-on to heads-on, from manipulation to mathematisation; (6) aiming for deeper and higher-order thinking and reasoning; (7) teaching towards 'learning to learn' with good habits; (8) cultivating positive attitudes towards mathematics and peers. These common themes fall into three overarching domains: (1) reasoning about mathematics, teaching and learning; (2) great emphases on learners' cognitive, emotional and social development past, present and future; (3) constant observations and diagnoses of learning.

In the following sections of the chapter, we first give an overarching discussion on teacher knowledge, beliefs and self-efficacy, drawing on both the teacher interview and survey data. Then, detailed description will be given regarding the eight themes emerging in the teacher interviews. Finally, we integrate all the findings upon closing the chapter.

8.1 Teacher knowledge, beliefs and self-efficacy

When interviewed immediately after their lessons being delivered and observed, teachers all start with acknowledging the location of the content in the textbook, before reflecting on the lesson design, implementation and their teaching beliefs in general. In comparison with findings of similar interviews with Chinese maths teachers from our previous project (Miao & Reynolds, 2018), much deeper pedagogical thinking and reasoning about the content and students is evident. Throughout the interviews, none of the teachers talks solely about the mathematical content or students for a length of time. Teachers articulate their thoughts and reflection by weaving the two together all the time. However, we do realise that underneath the tightly woven thoughts are three key strands of teacher knowledge: knowledge about mathematics (topics and connections between them), knowledge about the learners (their backgrounds, experience, prior knowledge, attitudes and habits) and knowledge about mathematics learning (both basics and higher-order).

The teacher questionnaire offers some explanations. All 70 teachers had attended initial teacher education (ITE) programmes in formal education (师范教育). Their ITE all included the three elements listed in the TALIS 2013 item TT2G12 – content of mathematics, pedagogy of mathematics and mathematics teaching practice (OECD, 2013). Such a high proportion, however, is not typical for this particular group of teachers but for most teachers in China where teacher training institutions are traditionally set to educate subject specialists in three- to four-year programmes.

In the teacher questionnaire, we also asked teachers about their beliefs and self-efficacy, using two scales (TT2M14; TT2M15) adapted from the TALIS 2013 (OECD, 2013).

About their beliefs, all teachers hope to develop the structured and logical knowledge of mathematics amongst their students (70/70, item 2) and believe that developing logical thinking amongst learners is one of the missions for mathematics teaching (70/70, item 9). They all agree or strongly agree that getting a correct answer and rationales behind the answer are equally important (70/70, item 3). Almost all of them believe that thinking creatively and making hypotheses and estimation are necessary for mathematical practice (69/70, item 10) and that basic skills come before complex problem solving for learners in mathematics (68/70, item 6). They expect students to make independent judgement and evaluation of their solutions to mathematics problems (64/70, item 5). The majority of the teachers disagree that memorisation can be used to learn most content in mathematics (62/70, item 11). Most of them agree that the purpose they teach is to facilitate students to tackle real-world problems with mathematics (59/70, item 1); the reason the other 11 teachers choose to disagree is likely because they do not take this as the entire goal of mathematics teaching. Only about two-thirds of the teachers prefer to have students tackle several hard problems than many easy ones (47/70, item 8). Less than a half of the teachers believe that problem-solving competence can be enhanced by asking students to solve difficult problems in the class (30/70, item 7).

In terms of self-efficacy, all or most teachers think they are able to nurture mathematics confidence amongst their students (70/70, item 5). They believe that they are capable of facilitating deep thinking in their class through posing thought-provoking questions (69/70, item 1) and that they are well aware of student status of understanding (65/70, item 3). Being confident in getting students interested in mathematics (65/70, item 2), most of them believe that their teaching makes students see the fundamental concepts in mathematics (64/70, item 6). Generally being confident about their teaching efficacy, the proportion of teachers finding it easy to meet individual students' needs drops to 40 out of 70 (item 4).

With the capsuled information on teacher knowledge, beliefs and self-efficacy in mind, let us now see them in a more vivid picture defined with thicker colour and richer details.

8.2   Teachers as living synthesised textbooks

Chinese teachers have a tradition of systematically studying the teaching materials including curricular standards and textbooks (Li, 2004).

Our interviews with the master teachers are unstructured, but they all kickstart the conversation by talking about the lesson content. Teachers all know the curriculum standards and textbooks in detail by heart, demonstrating a robust understanding of mathematics as a system. They would refer to a specific task on a particular page in a unit, without any forms of concrete material in hand. It is all safely stored in the mind. Not just that, the most impressive part is their robust understanding of the mathematical content of all phases at both macro and micro levels. Talking about the content they teach during the post-lesson interview, almost all of them introduce the lesson by clarifying the particular location of the content in the primary curriculum and textbooks. For example, Mr X_{109} starts his reflection on the lesson with reference to the location of the content: '*Su Jiao Ban* (苏教版, meaning 'Jiangsu version of textbooks', PhEP, 2014c), Unit 4, *Definition and Meaning of Fractions*'.

Perhaps for any teacher in a post-lesson interview, the first instinct would be referring to the content. However, the aspect that impresses us most is that they do not just talk about what is covered in the lesson but explain explicitly where the content stands in school mathematics, and not just in primary mathematics but very often the connection runs into secondary. Mr W_{104}'s school is catering for children in Grades 1–9 covering the compulsory phase of education – primary and lower secondary. Though responsible for primary mathematics, he maintains an interest in studying both primary and secondary curricula and textbooks and observing lessons in the secondary grades as well as the primary grades. He is looking for chronological trends of learning as it takes place.

If being familiar with the curriculum and textbooks is a threshold for those in the profession, then being a master teacher means robust understanding of and strong competence to synthesise the materials and learner status in all important dimensions – cognitive, emotional, metacognitive, physical, psychological and social. At the beginning of the interview, without the textbook in hand, Ms C_{512} starts her reflection by introducing the content of the lesson: the worked problem #1 on page 100 in Primary Maths 6B published by the People's Education Press (PEP, 2014b). Then, the conversation starts with her explanation about the actual unfolding of this particular problem which apparently plays a key role in the lesson.

She explains her rationale for cutting into the mathematical pattern underpinning this entire category of problems by starting with 20 points instead of 8, as readily given in this worked problem #1 in the textbook:

This is about mathematics reasoning in the case of drawing line segments using given points. In the textbook, the number of points is 8. I changed the number slightly greater to 20 as a starting task. Since the number is greater, the class are challenged to take action in seeking a simpler solution for the problem. Of course, I started the lesson with a historical story to remind them of the mathematical thinking strategy – 'making the complicated simpler (化繁为简)'. I used the ancient story, *Cao Chong Weighing the Elephant*. We have used this strategy previously in Grade 5 when learning to calculate the volume of irregular shapes.

(*MasterMT, 20190419, itv512*)

This ancient story taking place about 1800 years ago is about Cao Chong, the son of the Minister of the Han Dynasty, Cao Cao. One day, his father received an elephant as a gift from a general from the Kingdom of Wu. Cao Cao wanted to know the weight of the elephant, but nobody knew how to weigh it properly, except the six-year-old Cao Chong. He asked people to put the elephant on a boat in the lake and marked the level of the water on the side of the boat. Then, the elephant was led offboard and replaced with many rocks that submerged the boat to the marked level in water. He then asked people to remove all the rocks from the boat and weigh them. The elephant's weight was thus transformed into the total weight of all the rocks on the boat.

This story is included in Chinese literacy textbooks for primary students. Everyone knows it. Ms C_{512} uses the story as a classic example to show vividly students a method of mathematical reasoning that is widely known and well regarded in China: *making the complicated simpler*. This may include the case of transforming a problem with greater numbers to a problem with smaller numbers such that one could easily get the basic mathematical pattern. In this case, she hopes to nudge the class to start their exploration into the pattern with a number smaller than 20, by initially asking them to work with 20 points. If she chooses to use the 8-point situation in the textbook, the students might not see the number 8 as an obstacle and may thus go directly to finding solutions by connecting points. That way, they will not see the need to be strategic and the necessity to replace the bigger number with a smaller one, hence *making the complicated simpler*.

In the lesson, with the 20-point problem on the screen, Ms C_{512} initiates a conversation in the class in an attempt to anchor the exploration to a smaller number suggested by the students:

Ms C_{512}:	Let's see how to connect all the points.
	[clicking the remote control to quickly show the animated drawing of several line segments (see Figure 8.1b)]
	If we draw the line segments like this, what do you think?
Students [Ss]:	It is chaotic.
Ms C_{512}:	It is chaotic. Spot on. It's a bit messy. It's easy to repeat or . . .
	[looking at the class and waiting for them to complete her line]
Ss:	. . . omit [counting].
	[completing Ms Chen's line]
Ms C_{512}:	Isn't it? So, we have to think of a good method.
	[pausing for 2 seconds]
	What number should we start with to study (the problem)?
	[pausing for 2 seconds; some students raising their hands]
	Okay, tell us your idea.
	[pointing to student 1, S1]
S1:	We can start from four points to study it.
Ms C_{512}:	Ah, four points. Some say three points. S2?

每2个点能连成1条线段，20个点能连成多少条线段？

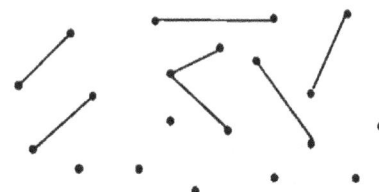

每2个点能连成1条线段，20个点能连成多少条线段？

Translation: A line segment can be drawn between two points. How many line segments can be drawn if there are 20 points?

操作要求：
（1）先画图连线
（2）再想 · 想你是如何连线的？能否用 · 个算式表示出你的结果？

Task Requirements:

(1) Draw the line segments first.
(2) Think: How did you draw them? Can you use an arithmetic expression to show your results?

FIGURE 8.1   The leading-in problem shown on screen in Ms C_{512}'s class

S2:	Five points.
Ms C_{512}:	Ah, five points. Aren't these all suitable?
	[looking around the class]
	What is the method that they all used?
Ss:	Transforming a hard task into an easy one (化难为易).
Ms C_{512}:	Exactly, transforming hard tasks into easy ones.
	[pointing at the slide]
	In other words, turn the large number into a small number, or turn the complex task into a simple one. So, let's start with five points, okay?

She then reminds the class to look at the task requirements written on the work-sheet (Figure 8.1c) before they start to work on the simplified 5-point problem independently.

Teachers not just know the content of the lesson they delivered. They know them in-depth and store them at the back of their mind as part of the 'skeleton' of their knowledge, making free retrieval of both small details and the grand structure of that knowledge easy. They reason rationally about the organisation and function of knowledge along the timeline. After the lesson on *Enlargement and Shrinking of Shapes*, Ms L_{508} talks about her view of the content, having studied the textbooks thoroughly:

> It is very important to understand the textbooks holistically. Though this lesson content is organised in the unit on proportionality, *Enlargement and Shrinking of Shapes*, together with *Transition of Shapes*, *Rotation of Shapes* and *Axisymmetric Shapes*, all belongs to *Movement of Shapes*. In fact, it plays an essential role in knowing and measuring shapes which may become simpler because of it. [In the textbook,] there is nothing else other than proportionality that is closely related to it. I reckon it is a preparation for learning *Similar Shapes* in the secondary school stage.
>
> (*MasterMT, 20190417, itv508*)

Ms L_{508} acknowledges that *understanding the textbooks as a whole* is the major focus of the teaching research project that runs across her district. For her, this is a goal that must be realised. Her teaching shows that the holistic approach to using textbooks is indeed carefully implemented in her classroom.

Our impression is that the master teachers have indeed thoroughly mastered the content and can refer to any tiny detail or big chunk of knowledge instantly during a conversation and most importantly during teaching. Their subsequent explanation of the content as part of a knowledge system shows that they are truly living textbooks with insightful syntheses.

8.3   Scaffolding the learning process

Another theme on the beliefs of master teachers is that they all pay serious attention to the process of learning. To give students the best experience of learning in the class, they design the process carefully with all the necessary scaffolds readily planned and flexibly delivered, by teaching and reflecting on the feet (Schön, 1983). Thinking 'on their feet' as they teach is evident in the interview where they voluntarily talked about how and why they planned and implemented particular steps in the class.

In the Grade 3 lesson on *Strategies for Problem Solving* (PhEP, 2014a, p. 27 in Unit 3), Mr B_{303} sees the importance of getting the students to think about the solution by looking at both the question and the given conditions in a word problem.

> Children tend to delve straight into the problem solving, before thoroughly analysing the problem. Of course, the analysis and problem solving are often interwoven. But students tend to jot down a number sentence immediately and solve it without carefully considering the complete information. So, I expect the students to not only analyse the problem, but also write down the analyses in simple words. I ask them to analyse not only the question but also the known

information from which they must select the information that is useful. This way, students are able to think from two directions – starting from the question and starting from the given information – and finally solve the problem. These are in fact the so-called 'synthesising method' and 'analysing method' respectively. With kids, we don't name the two methods formally as such, though.

(MasterMT, 20190514a, itv303)

Similarly, to facilitate children's thinking about how to properly enlarge or shrink a shape, Ms L_{508} asks the class to compare the perimeters, areas and angles of a shape against an enlarged/shrunk version. Thinking around these issues would lead the class to a possible discovery of two conditions that make a shape properly enlarged or shrunk: (1) the ratio of corresponding sides is the same; (2) the size of corresponding angles is equal. As they proceed, students can see the two conditions that prevent the layout of the shapes from changing as their sizes change. The change in size is reflected in the change of length of corresponding sides which in turn results in the change of perimeter and area.

8.4 A connected open system of knowledge and knowledge about knowledge

During the interview on revision lessons, teachers talk about the importance of revision and their systematic plans in making a revision lesson engaging, insightful and impactful for the learning mind. It is because of systematic planning that all the revision lessons that we observe are so interesting and engaging.

When teachers talk about a specific area of mathematics, they always put it in the broader system of mathematics rather than seeing it as a fragmented topic. For example, reflecting upon the grand revision lesson on *Equalities and Equations*, the teacher Mr W_{306}, says that during the planning stage he reflected upon the meaning behind the existence of algebraic expressions:

Whether or not using alphabets in expressions and equations sets the boundary between arithmetic and algebra.

(MasterMT, 20220515a, itv306)

In the revision lesson on the unit, *Definition and Properties of Fractions* (PhEP, 2014c, pp. 75–76 in Unit 4), Ms W_{311} aims to help students construct the map of knowledge in which various knowledge points mutually connect. In her view, students' understanding should be built on the connection between different points/ parts of knowledge. About revision lessons, what she seeks to develop amongst her students are profounder understandings, broader visions and deeper thinking about the knowledge learnt:

Textbooks are compiled with its own system where various points of knowledge are sequenced in a particular way. So, at the end of each unit, after the students have thoroughly reflected upon what has been learnt, I always ask them to think carefully why these points of knowledge are learnt in such a sequence. In fact, this thinking process is to get them see the connections between various points of knowledge and then build the connections using their own methods – methods that may sometimes sound immature.

(MasterMT, 20190520, itv311)

Across classrooms, the final product of teaching seems to be learners mastering a connected system of knowledge and being able to apply the knowledge – just like the 'fraction masters' from Ms Q's class or the 'problem-writers' from Mr B_{303}'s or Ms T_{416}'s class who pose problems of a very fine quality. This is well articulated by Ms Z_{314} with the metaphors she uses to describe the purpose of her revision lesson on 3D shapes:

forming knowledge points into lines and then nets and planes. The ultimate goal is to enable students to see knowledge as not only trees but also forests.

(MasterMT, 20190522, itv314)

By inviting students to talk about the deducting process of the surface area formula and the volume formula of each 3D shape that has been learnt, her purpose is to further ask them to build maps that connect these formulas and then apply them flexibly to real-world problems.

Revision is an opportunity to look back, but teachers often show their intention to prepare students for future learning. Like the key word Ms Z_{314} writes on the knowledge map on the chalkboard (Figure 4.14e), growing (生长) is the status of one's knowledge (or knowing). Such a forward-looking mindset is a feature across interviews with teachers. Teachers all talk about students' prior, present and future learning, looking back to the design and implementation of the observed lessons during the post-lesson interview. They seem to naturally take a chronological responsibility for students' learning, though nobody asks them to. It seems as if, to them, teaching is not properly done if it does not have a lasting effect on learning.

The design and implementation of each lesson is thus guided by an integration of (1) the past, present and future of learner development, (2) learners as individuals and a learning community and (3) learning in (meta)cognitive and socioemotional dimensions. In thinking and doing so, teachers practise metacognition constantly as they think about other's cognitive status, development and self-regulation.

8.5 Cultivating thorough understanding amongst students

The master teachers seem determined to get the students to understand both the concepts and procedures at a profound level. In their view, knowing the *what* and *how* and knowing the *why* are equally important. It is not a dichotomy. It is about both.

Reflecting upon her lesson on *Multiplying a Fraction by a Fraction* for a class of fifth graders, Ms L$_{410}$ explains the main aim of the lesson:

> Often many children know the method for calculation but they do not understand the rationale behind the calculation. Therefore, the major aim of this lesson is to study the rationale. In the lesson, I allocated a relatively long period of time for the class to think about how they come up with the answer for $\frac{1}{2} \times \frac{1}{5}$ and to represent their justification of the calculation they have made. Students are invited to talk about and represent their methods.
>
> (*MasterMT, 20190531, itv410*)

Similarly, in the Grade-5 lesson on *Adding and Subtracting Fractions with Unlike Denominators*, Ms L$_{304}$ considers the key to a deep understanding of the calculation is understanding the 'rationale behind calculation', a term many teachers have explicitly talked about throughout their interviews.

> Simply being able to do the calculation is not enough. The most important thing is to get the students understand the rationale behind calculation. In this case, the rationale for converting fractions with unlike denominators into fractions with like denominators before adding or subtracting them is similar to addition and subtraction with whole numbers or decimals: they must have same units in order to calculate. Fractions with like denominators have the same unit fraction. Children's learning experiences must be enriched over the course of their sharing ideas openly at both the group and class levels.
>
> (*MasterMT, 20190514, itv304*)

To be able to cultivate profound understanding, teachers need to have an accurate diagnosis of students' status of knowledge: What have they learnt? What are the aspects that might pose a particular challenge? With accurate answers to questions like these, teachers make accurate prescription before the lesson and then optimal adaption in the class.

Reflecting on the introductory lesson on *Trapezoids* for fourth graders, Ms P$_{315}$ first talks about the prior knowledge and then discusses one of the crucial steps in the lesson:

> In terms of cognition, students have already known basic properties of triangles and parallelograms. The lesson is therefore based on students' experiences in mathematics learning and life. I thus start the lesson by asking the class from which aspects they think trapezoids can be 'studied'. Students gradually get to know more about trapezoids in the course of attempting to draw one. When errors in drawing the shape occur in the class – for example, when a student draws a parallelogram instead of a trapezoid, I seek the opportunity to ask the

class to compare the two kinds of shapes. The comparison helps the class understand the essential differences between trapezoids and parallelograms. This way, the fundamental properties of trapezoids are highlighted.

(MasterMT, 20190522a, itv315)

With deeper understanding of both students and mathematics, teachers press for explanations, justifications and connections when correct answers have appeared in the class or when students seem to know what it is. They seek this opportunity to further prompt for the rationale behind the mathematical facts or fundamental properties underpinning certain concepts right after the facts or concepts emerge from the class.

8.6   From hands-on to heads-on, from manipulation to mathematisation

For teachers like Mr S_{201} and Ms Y_{202}, the process of getting to know the Möbius band must be fun, hence involving opportunities of papercutting, the exciting a-ha moments, and the joy of presenting the completed work to peers. But these are not enough. They are just halfway through the learning path of this lesson. The next half is to transform the hands-on experience to heads-on understanding such that students can go beyond the manipulation of paper stripes, do spatial reasoning in an abstract manner, and mathematise the mathematics beneath those interesting phenomena. In the interview, Mr S_{201} talks about the deeper outcome he aims at:

The textbook gives more emphasis on the hands-on activity of paper cutting. However, I think if the lesson only ended at cutting paper, students would not be able to learn much. They should be able to find the pattern beyond the cutting, particularly in the case of cutting the stripe into three, four or five equal sections. These are not shown as figures in the textbook. Is it necessary to carry on cutting the paper or is it time to seek insight from the cutting processes? . . .

I think the key in this lesson is to facilitate students to think and talk about the phenomena. So, the entire lesson is designed for children to explore from simple to complex situations and experience the process of questioning, thinking/hypothesising, testing hypotheses through hands-on activities, analysing the rationale, and then rethinking. Of all these, the key is thinking.

(MasterMT, 20190422, itv201)

Acknowledging that the content is in fact from Topology, one of the most challenging parts of college mathematics, Mr S_{201} says that this lesson is the type of content labelled as *Maths is Fun* in the textbook. Nevertheless, he believes what children should get from it is the mathematical methods underneath the fun of learning. What he plans to do in the next lesson will involve thinking about and comparing the situations where the stripe is cut into an odd vs even number of equal sections. By doing so, he hopes that children can carry out mathematisation.

Also having taught a lesson on the Möbius band, Ms Y_{202} starts the post-lesson interview by reflecting on the expectations of the textbook and national curriculum standards. She points out that, according to these, students are expected to experience the magic of the Möbius band and find the learning process fun. However, she thinks having fun is not enough:

> I hope they can get more out of it. The process of hands-on and reasoning about the possible result of papercutting develops students' spatial competence . . . . It is important to remind them to have a clear purpose before actually taking actions to cut. In addition to reasoning, spatial competence, and having purposes, they have gone ahead, analysing and talking about it, which is important.
>
> *(MasterMT, 20190422, itv202)*

Underneath such beliefs and teaching arrangements of 'from hands-on to heads-on' is the purpose to foster deeper and higher-order thinking and reasoning in and about mathematics. The learning process should be fun, which is not enough in and of itself. Doing mathematics is superficial without seeing the mathematical patterns and rationale behind all the manipulation.

8.7 Aiming for deeper and higher-order thinking and reasoning

Deep down, master teachers all appreciate the kind of higher-order competencies that could last long. It is evident across classrooms that master teachers have put such value into practice, seeking to cultivate higher-order competencies amongst their students. The interviews with teachers confirm that the emphasis on higher-order competencies in the class is rooted deeply in their belief systems.

Mr W_{211} talks about the lesson on The Highest Common Factor (HCF): what I hope to do is to shed new light on 'old' content that has been considered difficult to teach in all editions of textbooks. He says that the curriculum demand has changed from *finding* the HCF in the previous version of curriculum standards (MoE China, 2001) to *looking for* the HCF in the current standards (MoE China, 2011), with the emphasis placed on knowledge gained over the *process*. Between summative and formative aims, he seeks to find an optimal balance, free from the 'either-or' trap:

> This is based on the core values of the latest curriculum standards – putting the children at the centre and offering them opportunities to experience the process of knowledge construction. However, looking for the HCF is too simple for our students. I need to maintain a balance between the two ends. To reach this overarching aim, I designed three activities. The first activity aims to activate students' prior knowledge about factors and multiples over the course of looking for the factors of two numbers in a rectangle. The second

activity transits smoothly to looking for the common factor (CF) in a square. The two activities allow the children to construct a good understanding of the concepts and over the course to think about the lowest and highest common factors. We are thus able to cut through multiple methods in the third activity where children also raised methods that had not been taught. For example, they have learnt elsewhere how to do prime factorisation and find the common factors using short division (短除法) which are beyond their level of understanding. With such a 'nonzero' starting point in learning, we must not avoid the fact, but I take the caution to get them thinking about the 'why' when they might have known the 'what' – you must know why your methods work. I am delighted to see that students have posed a new idea as to whether the difference between two numbers is their HCF. In fact, the difference is a multiple of their HCF.

The most invaluable reward is that children managed to pose new problems. The curriculum standards demand 'four abilities' around problems: identifying, posing, analysing and solving problems. In such a lesson that is not heavily problem-based, problems, as a driving source, generate new ideas. This is more meaningful to their learning in the long run than only getting to know how to find the HCF. In the end, I hope the visualisation of numbers (i.e., in shapes) can further help them build connections between the concepts of CF and HCF. This will enable the children to see all sorts of connections between the seemingly boring and abstract numbers. Mathematics is a system, and it is within such a system that knowing and understanding can be constructed.

In the end, we present the problems in tiling the floor and loading cube boxes into a cargo container, with a purpose to further facilitate children to see that the essence beneath these problems is in fact about finding the HCF of two or three numbers. Though I have not asked the children to actually calculate the results, our [TRG] team's belief is already realised – 'the lesson may come to an end, but the thinking continues'.

These are about the lesson. Beyond the lesson, our team have broad aims. As a team, we have been through a series of reforms. Previously, we were more test-oriented, but now we focus on the expectations of the national curriculum standards. To realise this, we have renewed our ideas. We see mathematics knowledge as the lower-level aims, mathematical methods resulted from knowledge as the mid-level targets, and mathematical thinking formulated via mathematical methods as the higher-level purposes. In the course of practice, I have proposed a slogan, 'Cultivate the mind with mathematics, and nurture the heart with warmth.' Ultimately, we hope to make mathematics a warm subject.

(*MasterMT, 20190507, itv211*)

During the interview, two things are explicitly discussed by all teachers: mathematics and students. Teachers often start with their analysis of the location of the content in primary mathematics and even school mathematics across phases. Then, they all move naturally towards the status of the learners: what they have learnt previously, will learn in today's lesson, and will learn at certain points in the future.

In Ms W_{207}'s lesson, by asking the students to systematically sort the knowledge, she hopes to enrich students' knowledge about 3D shapes and develop their spatial competence. What she ultimately aims for in a revision lesson is to build overarching mathematical competences upon what the class previously learnt rather than repeating what has already been learnt and known.

> The difficulties are not only developing students' knowledge from 2D to 3D but most importantly from 3D backwards to 2D. After all, the rationale for calculating a shape's surface area or volume is rooted in its properties.
>
> *(MasterMT, 20190424, itv207)*

She reckons that in given problem situations, the first thing that occurs to students is to use a formula, but the rationale behind the formula is related to the shape's properties. These form the basis for this revision lesson. She comments on the revision notes that students did before the lesson. The major problem is that they all focus on specific points of knowledge but to a certain extent also neglect the connections between knowledge points at a deeper level.

After her lesson on *Introduction to Trapezoids* (PhEP, 2014b, Unit 7), Ms P_{315} continues her reflection on the lesson from where the content is located in the textbook and then along a broader learning timeline:

> Introduction to Trapezoids is the last session in this Unit on 2D Shapes, and it is also the start of the domain, *Space and Shapes*, for mid-graders. Before learning today's content, students have previously learnt about the properties of triangles and parallelograms. So, I designed today's lesson based on their learning of triangles and parallelograms. It is built on their experiences in both learning and life. Thus, I started the lesson by asking them from what aspects they thought the trapezoids could be studied. This way the children had an initial plan for studying the topic. Then, through drawing trapezoids, students gradually deepen their perception of the shape. A student drew a parallelogram by mistake. This offered an opportunity for the class to compare the two shapes. The comparison showed the difference between parallelograms and trapezoids, which thus revealed clearly the fundamental properties of trapezoids. It is important to help students to construct their knowledge with a logical structure. To study shapes, I think it is very important to let students form points of knowledge into a system. Within that system, they are able to think about questions as to what methods can be used to study a shape and which properties of the shape can be studied.
>
> *(MasterMT, 20190522b, itv315)*

The teacher also arranged a hands-on activity for students to fold and cut a trapezoid from a piece of paper. What she planned to do was to ask the children to create a trapezoid using different shapes that they had learned before, such as triangles, squares, rectangles and parallelograms. The mathematical experience that the lesson could offer is in line with the *four basics* in the new curriculum standards (2011):

> I think in addition to the basic knowledge and skills (the traditional two basics in the curriculum standards), the emphasis must be put on basic mathematical activity and basic mathematical thinking (the newly added two basics). This is what I pay particular attention to in teaching.
>
> *(MasterMT, 20190522c, itv315)*

With these in mind, Ms P_{315} builds the overarching outline of the lesson which should start from real objects in life, arrive at the focal shape in an abstract sense and then return to concrete manipulation of the shape. In so doing, she hopes to facilitate one of the key competencies in primary mathematics curriculum – spatial thinking. Similarly, Ms L_{508} shares in the interview that her lesson on *Enlargement and Shrinking of Shapes* aims to contribute to the development of children's spatial literacy.

Considering the new expectations from the curricular reform since the turn of the new millennium, Mr X_{109} pays attention to incorporating key competencies into everyday classes, such as communication and collaboration. Acknowledging that verbal communication skills are important in subjects like the mother togue, Chinese, Mr X_{109} argues that such an ability is equally important in mathematics:

> In today's lesson, including the parts on the meaning of fractions, groupwork and class discussion, students were able to talk and make timely response to each other. These are in fact for them to practise their communication skills. Timely response and comments from the teacher can encourage them to talk more. Such skills will be very helpful when the students grow up and enter the society as adults.
>
> Another important competence is collaboration. In the 21st century, collaboration becomes increasingly important. Often it is a win-win situation. It is thus very important for people to form a team in the workplace. In today's lesson, I have organised collaboration in groups of two and four. Of course, these are organised in a specific sequence. Children get the opportunity to become aware of their roles in the groupwork, split the task between them, be a team player, practise listening skills with others, and mutually inspire each other as they brainstorm together. In fact, this is aimed for the social aspect of students' development . . . . Ultimately, I hope they could realise that there are goals that a team can realise whereas an individual cannot.
>
> *(MasterMT, 20190402a, itv109)*

One month before his students graduate from primary school, Mr W_{306} talks about the deeper purpose of his grand-review lesson for sixth graders on *Equalities and*

Equations. He explains that he asks the students to talk about the pros and cons of their peers' presentations so that they have the mindset to first appreciate and draw on others' strengths and then seek points for improvement. If students are able to, then the teacher would consider his wish for this semester has come true. Though emphasising solid understanding and practice of the content, the teacher aims ultimately for overarching competence:

> At the beginning of each lesson, there is an activity called My Gigantic Net where one to two students present their construction of [specific mathematical] knowledge. Each time, after the presentation, I ask the class two questions. On the one hand, I ask them to talk about what they have learnt from the presentation to improve themselves by learning from others. On the other hand, I ask them to offer suggestions to the presenter. This gives them opportunities to develop critical thinking skills.
>
> . . . The lesson must leave with the students the competence that they can take with them into lower secondary grades.
>
> . . . It's only one month to go before the primary phase ends. This 3-min presentation activity holds within it a relatively higher aim and expectation from me. I hope to use this as a sending-off gift for the students to carry into lower secondary schools. It will make them better prepared for the learning in secondary.
>
> (*MasterMT, 20190515b, itv306*)

When talking about the purpose of lessons as a whole, or a small event in a specific moment of a lesson, like Mr W_{306}, teachers all emphasise learners' cognitive, emotional and social development in the past, present and future. They view learning as a connected whole both horizontally (connected to other topics, domains and even subjects) and vertically (connected to learning at various points in time). Teaching underpinned by these considerations develops students' higher-order competencies on a daily basis. It paves the path that leads to learners knowing how to learn.

8.8   Teaching towards learning to learn with good habits

A common feature arising in the interviews is that teachers show a strong determination to cultivate metacognitively competent learners who know how to learn and have good learning habits.

Mr B_{303} has been a nationally renowned Super Teacher with a Professoriate-Senior title for many years. He is one of the few 'celerity' teachers who are still teaching on a daily basis in average classrooms. About the lesson observed, he recalls that two students were randomly selected to come forward and share their solutions via the projector first. They happened to be both wrong, which was not planned. They were selected as a result of drawing straws from the name box. This is part of the class's convention – everybody gets a chance and it is completely random. The teacher thinks that it is always good to have students talking about where they went wrong. Without seeing a rich collection of correct and wrong solutions, students cannot learn to learn.

> All the students would think they were right when they wrote down their solutions. If they knew they were wrong they would not refuse to correct it. No one will say, 'I know I am wrong but I don't want to revise it.' When they are solving it, they all think they are right. So, asking the children to check their own solutions is a real challenge for them. When they are able to understand that they are wrong, they are no longer of the level where they write down the initial solution. They must think that they are right initially. It is by talking others through what they think that they are able to realise that something went wrong. This process sees them replacing one idea with another, which we conventionally call correction of errors. But correction is not easily done. It is not that simple. The process of real correction must be a process where one idea is replaced by another. Only after being able to see what went wrong in their original solution and how to make it right can children truly transform their thinking from wrong to correct. In our class, not only correct solutions are shared but also incorrect ones. Children are open to talk about their errors, '*Here's what I did previously. Here's how I do it now* [italics added by the authors].' They are used to it. We have been doing this for a very long time. If we give it a fancy name, we might call it a kind of 'class culture'. In the class-level discussion, everyone can join in, adding something or raising questions. As their thoughts echo and clash, their expressions of ideas become clearer. Being able to express one's idea clearly is also very important. What we as teachers need to do is to make connections between their thoughts. It is when their thoughts transform from the previous version [to the current version] that *true* learning happens. We always say that we must let learning happen, but *without changing thinking, learning can never happen*. Such change results in communications with others. . . . *Children must see multiple ideas and find similarities, differences and connections in them. It is only when they have an ability to do so can we say that they have learnt how to learn*. We often say learning to learn, but I think this shall not only be an empty slogan. In the learning process, students must get to compare different ideas and build connections between them. Then can they arrive at new understanding. I hope my teaching can help children gradually build learning methods and learning quality like this.
>
> (*MasterMT, 20190514b, itv303*)

Learning to learn is the process where children regulate their cognition to get some knowledge based on what they know. With such a learning mindset and good learning habits, students become strategic and self-regulated learners (Muijs & Bokhove, 2020; Schneider & Artelt, 2010) who are able to reach deeper, not just better, understanding of knowledge and skills.

8.9 Cultivating positive attitudes towards mathematics

Master teachers have one other common trait – they all want children to like mathematics. It is observable in their lessons that students are highly engaged in the learning and doing of mathematics both independently and collectively with peers. Post-lesson interviews shed light explicitly on teachers' mindsets of cultivating positive learning attitudes and climates in the class.

Near the end of the primary phases, Ms S_{502}'s class has their first revision lesson on *Numbers and Algebra*. After the lesson, Ms S_{502} reflects upon her intention to build fearless attitudes towards mathematics through positive teaching. Ms S_{502}, as the head of the Grade 6 TRG, says that she has dissuaded her team from making a revision lesson into one packed with a 'sea' of problems waiting to be solved one after another by the class. In her view, a revision lesson must:

> give the children something. Something new. Something that's not a simple repeat of all the things that they had gone through when they first learnt them. The main effort should be put into the enchancement of children's mathematical thinking, methods and problem-solving skills. So, the main focus in this lesson is twofold: knowledge organising and problem-solving skills. At this stage, children have all the knowledge. What they lack is the entire process of connecting the loose points of knowledge into lines and weaving them into a web.
>
> (*MasterMT, 20190410a, itv502*)

In the interview with the master maths teacher, even a metaphor is mathematical. However, it is not only about mathematics. It is, after all, about children learning mathematics, and she wants the children to have fun learning the subject. This is perhaps why she sets the task of debugging an email that's been bugged as the last task. In the 'bugged email', a student introduces school life in northeast China to a previous teacher. All the numbers in the email were mistakenly placed because of a 'virus' (Figure 8.2).

The entire class cannot help laughing seconds after the slide is shown. A student is invited to read the message out loud, and the class including the student himself keep laughing till the end. Then, of course, the message is quickly corrected via Q and A. Using this easy, humorous, yet very relevant task, the teacher wants the class to have the sense that mathematics does not have to be boring. Along with other engaging tasks, this bugged-email problem prevents the class from feeling bored, leaving them with positive attitudes to take away from a lesson that expects them to reflect upon all kinds of numbers that have been learnt over the past six years. The serious mission is taken in a not-so-serious yet effective way.

In terms of the cultivation of positive attitudes towards mathematics, Ms S_{502} emphasises the long-term goal that she seeks to achieve:

> It might be a goal the realisation of which I might not be able to see with my own eyes. Over the course of my interaction with parents, very often parents say that they were very afraid of mathematics back in school days and they blame

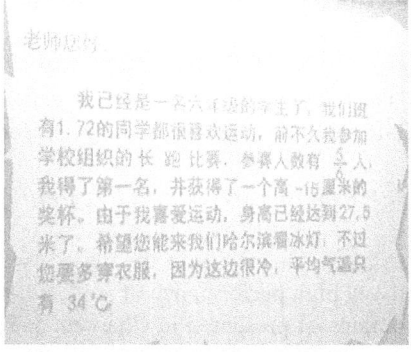

Dear Teacher,
I am already a sixth grader now. In our class, there are 1.72 students. We all love sports. Not long ago, I took part in a race at our school. Participants were 5/6 students. I got the first place and wan a cup that was -15cm high. Because I like sports, I am already 27.5 meters tall. I hope you could come and see the ice lanterns in Haerbin. If you do come, please wear more clothes. It's very cold here. The average temperature is 34 degrees Celsius.

Slide captured by the camera   *Translated from Chinese to English by authors*

FIGURE 8.2   The bugged-email task near the end of a revision lesson by Ms S$_{502}$

themselves for the fact that their kids struggle with maths. I hope by making the lessons fun and enjoyable these kids will one day become confident parents who are positive to mathematics and will pass positive attitudes on to their own kids in the future.

(MasterMT, 20190410b, itv502)

Similarly, one of the points that Mr X$_{109}$ makes explicitly is the connection between mathematics and the real world. In doing so, he wants to convey, as he always strives to do throughout his lessons, a deep interest in mathematics amongst the students:

Interest is the best teacher. Mathematics has something in itself seemingly boring, such as calculation and solving problems. Getting children interested in mathematics is what I have been trying to do in every lesson. Back to the lesson I taught today. Mathematics, from what I see, is never isolated from life – it is rooted in life. Last Friday, our school organised a spring trip. I took this as an opportunity to set the scene for today's lesson. I chose two pictures from that day. One captured a boy in our class, and the other captured all students from our school. I used the two pictures to showcase the whole can be one thing/person or one group of things/people. Judging by the implementation of the plan, I think these worked in engaging students from the beginning of the lesson. Why would they be interested? Because these were things that happened nearby. In fact, in all my lessons, there are always materials from the real world. I would like the children to know that mathematics is everywhere in the world where we live. All they need is a pair of sharp mathematical eyes [with which] they will see that life is full of mathematics.

(MasterMT, 20190402b, itv109)

However, with the positive attitudes come hard work, too. Mr W_{306} emphasises that students' mathematical confidence should be built on excellence:

> You cannot ask a student who makes mistakes every now and then or gets a 70% score for their work all the time to retain absolute confidence and yearn for mathematics learning.
>
> *(MasterMT, 20190515c, itv306)*

The positive attitudes and ability to learn are already a reality in the classes we observed, which is triangulated with the relatively high performance of students in the affective, metacognitive and cognitive domains as presented in Chapters 5–7. What teachers share with us in the interview offers us some of the explanation regarding the formation of positive attitudes towards mathematics.

8.10 Ready for evolving into master teachers?

In this chapter, we have listened to a rich collection of teachers' views regarding their teaching and their beliefs about what makes teaching and learning work.

Master maths teachers demonstrate an absolute mastery of mathematics both as a discipline and as a system of knowledge organised and defined in textbooks and curriculum. They have completely absorbed and internalised the mathematics, knowing by heart the location and standards of it in the teaching material, the connection within and between various topics, and the connection between mathematics and the real world.

They see their knowledge about the learners being as important as their knowledge about the mathematical content. To make the teaching and learning work, they task themselves with knowing both (knowledge about the learners and knowledge about the mathematical content), seeing the mission to best bridge the two as the fundamental purpose of any teaching. To be a best 'bridge' – to scaffold – they pose carefully designed key tasks, ask key questions and use key contributions (answers or work samples) from the students. Though these scaffolds sound intentional as they articulate them behind the scenes in the interviews, the various plans, when being carried out, look quintessentially natural in the class. This is because they are designed to look natural and work out just right.

Master maths teachers are not entirely satisfied if their students are only good at mathematics. Even just in mathematics, they want their students to be able to see the more profound meanings behind the abstract facts – they want students to see the logic, rationale, patterns, connections within and between the mathematical facts. They take every opportunity to get students to realise that life is full of mathematics and mathematics is both fun and relevant. They hope their students are happy learners who enjoy learning in general. Their lessons are designed with a long-term vision, such that students can learn to learn and manage both their learning and their life well, not only now but also in future. They aim to get students ready for future work and life – what they learn now should be still useful when they grow up. This kind of moral mission is also woven into their classes in which students are expected to understand and respect others and be able to collaborate. They design lessons with all these essential ingredients in mind.

This chapter has captured master teachers' current ideas and beliefs on mathematics learning and teaching. In the next chapter, we will travel back in time with the teachers to understand their journey of becoming master teachers.

9

GROWING INTO MASTER MATHEMATICS TEACHERS

Chapter overview

- What makes master mathematics teachers?
- Polishing lessons for public demonstrations and competitions
- Learning from and with peers in TRG and PD events
- Reading, reflecting, writing and publishing
- Learning from expert teachers and renowned master teachers
- Support from peers, school leaders and teaching research officials
- Studying the curriculum and textbooks in depth with great attention to lesson planning
- Researching as practitioners: a different kind of research
- Keep practising, keep changing
- Strong commitment to learner development and persistent interest in teaching
- Leading the way as a way of continuing learning and growing
- Master Teacher Studios: a cross-school professional learning community
- Participatory observation of teacher PD in action
- Observing a municipal teaching research conference
- Observing a provincial teaching research conference
- Observing a national teaching research conference
- It takes a village to nurture a master teacher

> By three methods we may learn wisdom: First, by reflection, which is noblest; Second, by imitation, which is easiest; and third by experience, which is the bitterest.
>
> – Confucius

DOI: 10.4324/9781003127925-9

This chapter presents two big pictures of mathematics teacher PD: (1) the longitudinal and rich picture of master teachers' PD trajectories over decades 'drawn' by master teachers themselves; and (2) the livestream picture of teacher PD in action at municipal, provincial and national levels drawn by us as researchers through participatory observation.

As introduced in Chapter 3, the second semester following our 'data-collection tour' in master teachers' classrooms, we decided to delve deeper into the cultivation of master mathematics teachers. To do so, we sent out a survey asking about teachers' PD trajectories and their MTSs should they be running one at the time of the survey. The survey provides us with the materials for the first big picture. We are then able to put together the second big picture with the data collected in three beyond-school teaching research conferences through participatory observation.

9.1   What makes master mathematics teachers?

The PD trajectories survey got a response rate of 38 out of 70; of those who responded, seven wrote about the development of MTSs and events led by them. The PD survey generated a total of 57,765 Chinese words written by the 38 teachers; the MTS survey response added up to 12,138 words. Together these teachers wrote 69,903 words sharing their career stories on PD and MTSs. These form the data on master teachers' PD trajectories and MTS-led PD. For the PD trajectories, teachers wrote on average 15% ($SD = 11\%$) on section 1 (bio), 58% ($SD = 24\%$) on the paths they had travelled, and 27% ($SD = 22\%$) on teaching research events that they took part in.

For the microanalysis, we looked at the word frequency of the 38 PD-trajectories documents and found 20 Chinese words appeared more than 100 times. These 20 Chinese words are in fact 19 by meaning, as two of them (老师 and 教师) both mean teachers. The analysis also identified the top three words mentioned by teachers more than 380 times: mathematics, teachers and teaching (Figure 9.1).

For the macroanalysis, we take the grounded theory approach where an open-coding strategy is taken (Glaser & Strauss, 1967). The open-coding strategy involves the initial coding of sentences and paragraphs by category and the constant grouping of similar categories into overarching themes. This results in ten major themes:

- Polishing lessons for public demonstration and competition;
- Learning from and with peers in TRG and PD events;
- Reading, reflecting, writing and publishing;
- Learning from expert teachers and renowned master teachers;
- Support from peers, school leaders and teaching research officials;
- Studying the curriculum and textbooks in depth with great attention to lesson planning;

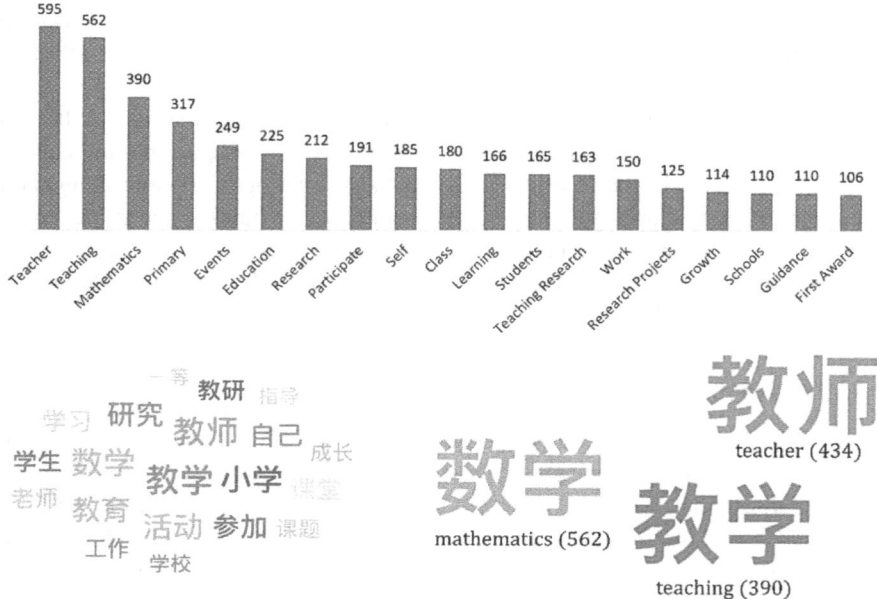

FIGURE 9.1 Words most frequently mentioned by master teachers looking back to their PD

- Researching as practitioners: a different kind of research
- Keep practising, keep changing;
- Strong commitment to learner development and persistent interest in teaching;
- Leading the way as a way of continuing learning and growing.

In the remainder of the chapter, we will first follow the master teachers through their PD trajectories to see what makes them master mathematics teachers and then see their PD in action by taking part in three teaching research conferences at the municipal, provincial and national levels.

9.2 Polishing lessons for public demonstrations and competitions

The most frequently mentioned contribution to teaching growth is the lesson polishing process as they prepare for a *demonstration lesson* (示范课), *open lesson* (公开课), *teaching-research lesson* (研究课) or *competition lesson* (教学比赛). Lesson polishing (磨课) has become a professional term mentioned by teachers when they reflect on their paths towards improved teaching. Throughout their career, these teachers seem to be tireless in presenting open or teaching-research lessons for teaching research meetings/conferences.

In the process of polishing a lesson for public demonstration or teaching competitions, their professional selves grow rapidly. During the early stage of his career, Mr C_{204} had been an active volunteer in delivering teaching-research lessons or demonstration lessons whenever possible. This is what he believes has led him to building solid basic skills for teaching. Mr C_{204} is not alone. All teachers write about the benefit of developing demonstration lessons and teaching-research lessons for teaching research events at various administration levels:

> Looking back at my development, I think that I benefited tremendously from the process of participating in various district- and municipal-level competitions. Each round of competitions offers me an opportunity to get help from colleagues and advice from experts, whilst striving to work at my best. It has also been very helpful to polish lessons before presenting them as teaching-research lessons or demonstration lessons at the district and municipal levels.
>
> *(MasterMT, PD-trajectories excerpt, Ms G_{209})*

> In the course of presenting demonstration lessons frequently, I have become steady on the stage. My teaching competence has been greatly improved.
>
> *(MasterMT, PD-trajectories excerpt, Ms W_{311})*

It is what they do in the process that transforms them into a more capable teacher. The process involves closely scrutinising every tiny detail of teaching and learning that has been systematically planned and carried out and that will continue to be. Video-recordings are used as a source for reflection and revision. Teachers are self-driven in the pursuit of a best lesson, although they clearly acknowledge that there is always room for improvement, hence no 'best' lessons:

> From lesson planning to making PowerPoint slides, from repetitive rehearsals to revision, in the month leading up to the competition, I often worked late into the night in the computer lab, which was well worth it. I had made a spurt of progress in ICT, whilst my teaching was improved over the course of countless rehearsals. When the lesson polishing reached its right moment, I made it by winning the first award in the municipal quality-teaching competition. . .. Since then, accumulating strengths has become the value-adding growth model in my teaching life.
>
> *(MasterMT, PD-trajectories excerpt, Ms W_{311})*

> I have been actively taking part in all sorts of demonstration lessons, open lessons and teaching-research lessons. To deliver a good lesson, I often record the lesson and frequently play back to look for points of improvement. I then redo the lesson and revise it on the basis of the recording. The revision won't stop

until a satisfying version emerges. For every demonstration lesson, I make all the teaching and learning tools by myself. Every occasion of lesson polishing is an opportunity to practise and grow. It is exhausting but rewarding.

(MasterMT, PD-trajectories excerpt, Ms Z$_{105}$)

Gaining skills and confidence through polishing and refining lessons, master mathematics teachers are ready to be 'fierce teaching athletes', pushing themselves to embrace challenges in all sorts of teaching competitions. There are major teaching competitions each year at the district, municipal, provincial and national levels. Candidates are recommended and selected by teaching research groups from the bottom up. The superficial aim is of course, like in any other competitions, to win. This means the recommendation and selection must focus on the most capable and therefore most likely to win. All the master teachers have won awards in teaching competitions at various levels. What they emphasised is the process of polishing a lesson on the same content numerous times. It is through constant constructing, deconstructing and reconstructing of teaching that their capacity to teach is upgraded. The new will not come unless the old is gone. The experiences are so intense that they all remember them vividly. Over the course of polishing lessons as part of a team, their understanding of students, textbook content and teaching and learning is uplifted, and as Mr W$_{104}$ put it, their teaching thus undergoes the 'transition from quantity to quality'. By taking part in the competitions, they have the opportunity to gain not only awards but also extra growth.

Having won the first prize in the national teaching competition for primary mathematics in 2009, I then won the first prize in national teaching with ICT competition for primary and secondary teachers in 2010. These were all based on the foundation and aspiration that I accumulated in the process of polishing a lesson titled *Mathematics in Bicycles*. In 2011, 2013 and 2015, I took part in the Municipal Competition of Essential Teaching Skills for Primary Mathematics, the Municipal Competition on Deduction in Primary Mathematics, and the 24-Hour Municipal Teaching Challenge of Primary Mathematics respectively. I won the first place in each of the events. Over the course of these competitions, my understanding of textbooks, teaching and students has increased. This is a process of transition from quantity to quality.

(MasterMT, PD-trajectories excerpt, Mr W$_{104}$)

I think volunteering to undertake all kinds of teaching-research lessons and demonstration lessons is very important. On the one hand, in the process of planning the lesson, one needs to delve deep into the textbooks, look for materials, communicate with colleagues and seek advice from experts, which is a precious learning opportunity. On the other hand, the teaching process itself strengthens one's communication skills and flexibility.

(MasterMT, PD-trajectories excerpt, Mr Z$_{205}$)

And they are not working on their own. Those who actively support them include, but are not limited to, their colleagues in the teaching research group/department, their school leaders and district/municipal teaching research officials. Everyone's involvement covers every aspect of the focal lesson under improvement. This specific lesson is scrutinised by multiple eyes and minds until it reaches a satisfying version.

> All the members in our mathematics department work together to offer strategies for me to refine the lesson.
>
> *(MasterMT, PD-trajectories excerpt, Ms Q_{401})*

> In 1995, I took part in the Municipal Quality Mathematics Teaching Competition on behalf of the school and won the first prize, following which was a first prize in the Provincial Quality Mathematics Teaching Competition. I would never forget that night 25 years ago when the headteacher appeared at our doorstep only because she was worried about me – a teacher with less than two years' teaching experience about to take part in the municipal competition. That evening, Headteacher Ms L acted as my class, with the wardrobe as blackboard and sewing machine as the front desk. We polished each question that I should pose and each tone I should be using in the class, until midnight. It should be noted that it was the 1990s when few people from around here had telephones let alone mobile phones. Ms L managed to find where I lived by knocking on the door from house to house. I would never forget the deputy head, Ms R, either. Just before the night was out, she handmade the beautiful slides and a cuboid model with six faces that could be slid into its frame. It is worth mentioning that it was the era when we didn't have computers at home and slides were all handmade with pens and films. Everything could only be made from scratch by hands. It is all because of these seniors who had offered their help and guidance selflessly that one generation after another of younger teachers could grow rapidly.
>
> *(MasterMT, PD-trajectories excerpt, Ms T_{416})*

Choosing to deliver a demonstration or teaching research lesson often means to teach fearlessly and perfectly to a class of unfamiliar students in the professional eyes of tens of thousands. Hence, the need for systematic preparation and polishing with the support of a team before it is presented. Such collective pedagogical thinking does not only take place before the formal presentation of the lesson, it also happens afterwards. Demonstration lessons are interwoven with deep reflections. Such reflections are often written down and eventually get published. This form of professional writing is already woven into the teacher promotion system in the country.

> A teacher's best learning venue is the classroom. I have taught lots of demonstration lessons. . ... Every time I completed a demonstration lesson, Headteacher Tang would ask me to write it as a teaching case.
>
> *(MasterMT, PD-trajectories excerpt, Ms Z_{314})*

These teachers are willing to present lessons to peers, experts and the public and are looking forward to receiving feedback and improvement advice. Many teachers mentioned that they see teaching in public demonstration or competitions as a flash in the pan. The meaningful part of that is the process that leads to it, the improving process. Knowing that every day lessons may not have the time and resources to be brought up to the same level as demonstration lessons, they see each routine lesson as an opportunity to get closer to the quality that could be reached in demonstration lessons. As Ms W_{311} put it: 'Teach a daily lesson as if I am delivering a demonstration lesson.' Mr B_{303} made it even stricter: 'Teach the demonstration lesson as I do in my own classroom on a normal day; teach my daily lesson as if I were giving a public demonstration.' How self-driven! They see the various awards as a new chapter of growth. It seems as if, in their mind, teaching improvement would never come to an end. Instead of *teaching improvement*, in their mind, the central theme of practice should be expressed in progressive tense – *improving teaching*.

9.3   Learning from and with peers in TRG and PD events

Across master teachers' accounts of PD trajectories, there is a strong sense of being part of a professional learning community. Many attribute their improved understanding of teaching to peer observations and teaching research group meetings.

> I have been actively taking part in all kinds of training and teaching research activities. . . . Each year, I insist on observing 40 lessons delivered by colleagues.
> *(MasterMT, PD-trajectories excerpt, Ms Z_{105})*

> In my professional development, I think I benefit tremendously from the teaching research activities organised by the team at our school and from the collaboration, discussion and mutual learning with colleagues in our subject-based team. Though one person can run very fast, it is a group of people that can go far. One's own capacity after all has its limit. Our school's teaching research team is my powerful backing. Many of the wise and creative teaching ideas and highly effective teaching methods are inspired by our discussion in teaching research activities.
> *(MasterMT, PD-trajectories excerpt, Mr S_{201})*

> Joining the district level lesson preparation group, observing the teaching research lessons presented by the group members and doing my own teaching research lessons – these activities helped me tremendously.
> *(MasterMT, PD-trajectories excerpt, Ms S_{208})*

> I find the teaching research group in my school the greatest help. Immediately as I started off as a new teacher, I took part in all the weekly meetings of the

mathematics teaching research group specifically set for young teachers. I had learnt a great deal from the teaching research activities and maintained steady growth as a teacher over the course of learning.

(MasterMT, PD-trajectories excerpt, Ms S_{505})

In addition to the TRG activities within their home schools, master mathematics teachers also acknowledge the learning experiences gained in PD events organised outside the school, at various administrative levels.

Ever since 2001 when I took part in the provincial PD programme for Backbone Teachers, I had attended a number of high-quality lectures delivered by academics and experts and had the opportunity to exchange ideas with Backbone Teachers from across the province. These activities enriched my professional knowledge and left me with huge benefit.

(MasterMT, PD-trajectories excerpt, Ms L_{304})

I take opportunities to partake various kinds of research training and practice-based research, which becomes a main way for me to keep teaching and thinking. . .. In addition to taking the mandatory open courses for continuous education and attending the teaching research activities at the school, district and municipal levels, what I do is actively taking part in PD and research training programmes. The annual meeting of Master Teacher W's Work Station, the annual conference of Master Teacher H & Teaching Against Misconceptions, the national annual conference for New Century Primary Mathematics and the district PD programme for Backbone Teachers are the ones that I would definitely attend.

(MasterMT, PD-trajectories excerpt, Ms S_{203})

In comparison with their ordinary peers, the master teachers are keen to take part in lesson demonstration and teaching competition – a mindset self-trained from early on. Over the course of doing these, they got more opportunities to polish one lesson as many times as necessary with peers, teaching research officials and experts, learning from and with them. The TRGs offer teachers the readily available professional learning communities where teaching ideas are shared, polished and uplifted (Huang & Bao, 2006; Paine & Ma, 1993).

9.4   Reading, reflecting, writing and publishing

Teachers are regular readers of professional journals and books published on mathematics teaching or education. Many attribute their growth to reading, reflecting and writing. Finding time and space to think for themselves is crucial to them. Whilst deep reflection increases their self-awareness, many of them acknowledge that reading broadens their knowledge, resonates with their experiences in practice and thus generates deeper thinking and reflection.

> In work, I constantly expand my knowledge by persistently reading books about teaching and education and taking part in various kinds of learning and training programmes.
>
> *(MasterMT, PD-trajectories excerpt, Mr S_{201})*

> To become well developed, a teacher must be ready to delve deep into learning. The teacher must learn the basic knowledge and skills for teaching and education and then consolidate the basics. Learn the educational theories and teaching techniques and then strengthen the teaching skills. Study the textbooks intensively, read background materials, and read books on mathematics teaching methods, comparative education, and child psychology, so as to master the law of child cognitive development.
>
> *(MasterMT, PD-trajectories excerpt, Ms C_{512})*

> I persist on writing teaching reflection after each lesson, summing up experiences.
>
> *(MasterMT, PD-trajectories excerpt, Ms Z_{105})*

> I actively take part in all sorts of research projects which becomes a driving force for me to reflect upon my own teaching and improve my theory-based competence.
>
> *(MasterMT, PD-trajectories excerpt, Mr X_{109})*

> The foundation of the career is built on good accomplishment of everyday teaching. I strive to reflect carefully and jot down ideas in my teaching plan and explore relevant teaching issues with my fellow colleagues in my TRG. Doing so allows me to handle properly the essential and difficult points of the knowledge in teaching.
>
> *(MasterMT, PD-trajectories excerpt, Ms S_{201})*

> Whilst completing various tasks in work, I strive to find time to write in my teaching journal what I have in mind. Even if it were just a few dozen words, I would write them done right away. As widely known, reflection is one of teachers' competences. It tells whether or not a teacher is competent enough. It is only through persistent reflection that teachers are able to constantly improve themselves and grow.
>
> *(MasterMT, PD-trajectories excerpt, Ms S_{203})*

The growth of teachers = experience + reflection. Being diligent at reflection makes one's teaching more scientific, teaching behaviours wiser, and teaching effects approaching the ideal, which overall rewards the teacher with more credits. Through learning, researching and thinking, I am more cautious about my teaching behaviours: Did my previous lessons give more space for me to show off my 'teaching'? To what extent have I care about students' thinking? What kinds of questions work better in promoting the class to think?

(MasterMT, PD-trajectories excerpt, Ms C$_{310}$)

The deep reflection based on reading and practice in turn presses their urge to jot their thoughts down. These teachers have published a lot in professional journals, and some have even published books. For example, Mr B$_{303}$ is already a seasonal writer, publishing his reflections and thoughts on everyday teaching regularly. Getting published is partly due to the need for professional advancement, since one aspect of teacher evaluation is publication, and partly due to the fact that it helps them channel their thoughts out to their peers at large.

Every semester, to improve my professional and teaching competence, I strive to write a quality article on teaching that is of a satisfying quality.

(MasterMT, PD-trajectories excerpt, Ms Z$_{105}$)

In the process of research, I managed to gradually improve my competence in theories and educational research. Many of the articles that I wrote upon doing research have been recognised with awards from the district or higher-level educational authorities.

(MasterMT, PD-trajectories excerpt, Ms G$_{209}$)

Over the course of my exploration into these questions, I gradually published a number of articles in our provincial *Educational Research Journal*, including 'Thoughts and practice about behaviours in the mathematics class from the perspective of key competencies', and 'Towards harmony: The ultimate goal of modern classroom teaching'.

(MasterMT, PD-trajectories excerpt, Ms C$_{310}$)

To keep developing myself, I took time to learn theories, recorded the process of my practice, and tried to write about lesson cases and articles, which helped consolidate the teaching experiences that I accumulated.

(MasterMT, PD-trajectories excerpt, Mr C$_{204}$)

Because of the effort given to accumulating [thoughts and reflection], I published multiple articles out of it.

(MasterMT, PD-trajectories excerpt, Ms H$_{503}$)

As teachers reflect, read, write and eventually publish on their teaching practices, they know more about their professional selves, their teaching, their learners, their learning and mathematics as a system of knowledge and as a school subject.

9.5 Learning from expert teachers and renowned master teachers

All teachers have attributed their development to a number of key figures and their peers in the teaching research groups. Their gratitude was woven into the stories they told about the evolution of their teaching. At first, all teachers mentioned they started off by studying and mimicking video-recorded demonstration lessons delivered by renowned master teachers in the country whom they apparently regard as role models. The studying and mimicking processes sound relentless, hence demanding huge effort and great discipline.

All teachers write about the benefit of observing lessons, often in the format of video, delivered by established teachers or master teachers whom they admire enormously. Their learning from masterly teaching thus stems from the initial mimicking of others' teaching to the gradual development of their own styles.

> I have humbly asked advice from renowned teachers and studied lesson videos delivered by many master teachers.
>
> *(MasterMT, PD-trajectories excerpt, Ms H$_{503}$)*

> It is important to be ready to learn and to observe video-recorded lessons by those master teachers.
>
> *(MasterMT, PD-trajectories excerpt, Ms X$_{406}$)*

> I cherished the opportunities to attend the case study lectures by accomplished educational experts and scholars like Master Teachers, Q, H, W, S and Y. Their teaching styles – either steady, or humorous, or profound, or witty – struck me instantly, lingering in my mind forever. Ever since knowing that there were filmed distinctive lesson cases available for borrowing in the municipal E-Education Library, I had become a frequent visitor there and borrowed loads of videotaped lessons to study. I would play the lessons countless times to imitate. Then, everything comes to the one who waits. After tireless practice and polishing, I have formulated my own teaching style.
>
> *(MasterMT, PD-trajectories excerpt, Ms L$_{304}$)*

> I have always been looking up to the best in our profession as my role models. . . . One cannot improve professional competence by only looking at one's own practice. I therefore reached out to all potential opportunities and platforms to seek guidance from experts in the profession, which saved me from detours and helped me grow faster.
>
> *(MasterMT, PD-trajectories excerpt, Mr X$_{109}$)*

When I was a new teacher, I actively learned from experienced teachers, prepared lessons carefully, and observed colleagues' lessons. Through mimicking the teaching style of experienced teachers, I gradually accumulated my basic teaching skills.

(MasterMT, PD-trajectories excerpt, Mr C$_{204}$)

Not only had they learned from expert teachers' video-recorded lessons, many were lucky to have well-recognised colleagues or master teachers as their teaching *masters* (师傅, *shi fu*, as in Kungfu masters) appointed by the school as part of strategic support to novice teachers. Looking back, Mr C$_{204}$ attributes to his early growth to the learning experience with his mentors, Ms T in his home school and Ms W – the country's most widely known primary master maths teacher who is now leading the profession though not doing regular classroom teaching any more.

Three years after I worked as a teacher, the school appointed Ms T as my teaching master. Under the guidance from Ms T, I started to undertake district-level research projects and municipal-level teaching-research lessons. . .. With my teaching master's support, I had more opportunities to learn and achieved more. In addition, my teaching master had helped me reach out to more experts. It was during the process of doing teaching-research lesson projects that I got to know Ms W. This had helped me tremendously. With the support from the school, I became a member of Ms W's Master Teacher Studio where I got more opportunities to network with more accomplished experts and colleagues. With a platform like this, I was able to gain rapid growth.

(MasterMT, PD-trajectories excerpt, Mr C$_{204}$)

Learning from established master teachers and expert teachers can indeed speed up a novice teacher's development (Berliner, 2001; Leinhardt, 1989; Li et al., 2011). Such a novice-mentor system is universal in Chinese schools (Cravens & Wang, 2017; Fan et al., 2015; Huang et al., 2017; Li et al., 2011; Zhang et al., 2021).

9.6 Support from peers, school leaders and teaching research officials

These teachers do not just learn from their role models. They are ready to learn from anyone in the profession. From their point of view, another important source for development are those around them. The stories they told often happened in the teaching research groups as they were preparing a demonstration lesson for a teaching research conference or a teaching competition. The team often expanded to include temporary members, such as their headteachers and the teaching research officials from the local education authorities.

There is a readily available 'multilevel' system that supports teachers to continuously grow. These master teachers all made good use of the system from school up to municipal levels where teaching research officials treated them as 'seed' master teachers. Then, this seemingly hierarchical team worked together to improve the seed player's teaching practice. Very often they assisted the seed players to prepare, carry out and polish lessons for public demonstration or teaching competitions. The system is apparently available to everyone in the profession. However, this special class of teachers are those who, from earlier on, wanted to be better at teaching and who sought every opportunity to improve their practice.

All teachers wrote about their experiences of being part of the school-based teaching research groups. There is a mentor system available at the school level and sometimes at district, municipal and higher levels. Many teachers reflected upon their growth as guided by their teaching masters, as in Kungfu masters. They attributed their growth to having a good 'platform' where they could get access to quality PD resources.

> I was lucky to have the Director of Teaching at our school who acted as my mentor voluntarily observing my lesson every day at the beginning, then two to three times a week and then once a week.
>
> *(MasterMT, PD-trajectories excerpt, Ms G$_{209}$)*

> The 1997 Youth Cup Teaching Competition of the Xuanwu District was the first opportunity I encountered. It was later proved to be the very event that had triggered my teaching to go through a transformation in quality terms. Back then, working late at school was what I did on a daily basis. Besides the multiple teaching rehearsals, the head and deputy head of our school followed through my teaching and lesson explanations time and again. Within a very short period of time, I became more and more skilled at teaching. Living up to expectations, I got the first prize in the competition. This had given me the confidence and determination to strive forward.
>
> *(MasterMT, PD-trajectories excerpt, Ms L$_{304}$)*

> During the six years teaching at D School, I had received warm support from colleagues, school leaders and teaching officials from the district. I still remember the Springing Up Teaching Competition of the district which I took part in when I just started teaching there. To make my lesson more creative, all the teachers in our mathematics TRG worked together to mastermind teaching strategies for me. In the end, [in the competition,] this lesson left a very good impression to the then provincial teaching research official, Mr A, and the teaching research official of X district, Ms B. I had got lots of opportunities to challenge myself ever since. Every time I was about to join a competition, the district teaching research official, Ms B and Mr C would invite the members of the Primary Mathematics Centre to meet up to brainstorm for me. This team had brought me family-like warmth. About how to design and present every part of the lesson, everyone offered their ideas generously without reservation. When controversies arose in the discussion about the lesson, we would debate with

each other on teaching matters. There was no rank difference in the team; there was only light sparked, with different ideas clashing with each other. As team members, we have formed profound friendship. To make my lesson even better, the teaching research official, Ms B, and I often communicated online until late in the night about each line of the statement in the lesson and the design of the slides. It is because of the supportive team that I am able to successfully challenge myself and obtain a number of achievements and awards, including the first prize in the municipal teaching competition, the first prize in the provincial teaching competition, Master Teacher of the X District, the Subject Leader of the City, the Backbone Teacher of the Province, and the Subject Leader of the Province.

(MasterMT, PD-trajectories excerpt, Ms Q_{401})

Is the improvement of teaching an individual or collective endeavour? For master teachers, the seemingly either-or issue has reached a balance (Paine & Ma, 1993): teachers in the TRG have been working together to teach 'forward' independently. All the teachers looked for resources to improve practice not only from within schools but also from outside of schools. They have actively participated in PD events that not only give demonstration lessons but also include invited lectures by master teachers, textbook editors and scholars from higher education. In fact, many teachers emphasised the importance of reaching out and exchanging ideas with colleagues beyond their schools in other parts of the country. All in all, these broaden teachers' professional vision, making them constantly aware that teaching can be done differently and better.

9.7 Studying the curriculum and textbooks in depth with great attention to lesson planning

It is typical in China that teachers are generally required to study the curriculum and textbooks in great depth (Li, 2004). Amongst the MasterMT teachers, the extent to which attention was given to teaching materials was impressive. They had accumulated such deep understanding that textbook publishers reached out to them for assistance. Some of them were selected as implementers of a trial series of textbooks by textbook publishers in the hope that they could offer expert comments on the improvement of the trial versions. One of the MasterMT participants was actually a member of the author team for a textbook series. Another had been closely involved in the experiment of two series of new textbooks arising in the new millennium curriculum reform which transformed China's education from a system with one set of curriculum standards and one series of textbooks to a system with one set of curriculum standards but multiple series of textbooks. The curriculum reform commenced in 1999 and the new standards (experimental version) were officially released in 2001. Ms L_{304} had the experience of using the experiment versions of two textbook series: 'Modern Primary Mathematics (MPM)' and the series published by the Phoenix Education Publishing of Jiangsu Province (PhEP).

In 1999, I got invited to the experiment of the MPM textbook series. From grade one, I started to use the MPM textbook in one of the two classes I was then teaching, and the PhEP textbook in the other class. These two series of textbooks were different in many ways, in terms of organisation of contents and adaptability for teaching. I used the summer break to get myself into lesson preparation. The workload was huge for studying the compiling purposes and teaching content of two textbook series whilst designing the corresponding teaching methods. The process of comparison and reflection had developed my understanding of and ability to use the textbooks, uplifted my capacity to teach concisely and flexibly, and improved my mathematical competence. In the three years from 1999 to 2001, upon experimenting with the use of the two series of textbooks, I had delivered over 50 demonstration lessons, written multiple articles on teaching and reflection, and presented to fellow teachers in our district my analyses of the two textbook series and corresponding teaching research.

(*MasterMT, PD-trajectories excerpt, Ms* L_{304})

With the experience with two series of textbooks, Ms L_{304} had a positive expectation towards similar experiences. In 2001, the national-wide curriculum reform was formally implemented after the Curriculum Standards (experimental version) for Compulsory Education for all major subjects were officially issued and took effect, and she chose to embrace all the challenges and opportunities to grow as a teacher:

In 2001, the national new curriculum reform commenced, and the textbook series, PhEP, formally proceeded into its trial stage. I not only took part in the compiling and designing of the PhEP courseware for teachers and students but also actively participated in the experiments of the new textbook series.

This is a path less travelled and I had to use the textbooks one year before all other teachers started to use them. I must constantly summarise all the failures and successes of the experiment and learn from them. I was expected to share my teaching experience and thoughts with experts and teachers, which place a huge responsibility on my shoulder. In the six-year experiment of the textbooks, I had taken part in lots of national and provincial training events in my spare time. After carefully listening to experts' ideas and analyses about the textbooks and studying the reports and demonstration lessons based on the experiments that were carried out nationwide, I synthesised all the information with the characteristics of my students and my own teaching characteristics and tried my best to implement in my class the teaching concept embedded in the new curriculum standards, seeking to fully bring out the essence of the new textbooks.

Over those six years, I had written several dozens of lesson case studies which were published in the *Lesson Planning Handbook for Teachers*, offering guidance for teachers. I had thoroughly mastered the textbook and lesson content and permeated myself with the curricular spirit, which made me capable of teaching with skill and ease.

(MasterMT, PD-trajectories excerpt, Ms L_{304})

Similarly, for other teachers, it is more important than anything else to study closely the curricular standards and textbooks.

I highly recommend the new teachers to study *The New Curriculum Standards* and *Interpretation of the Curriculum Standards* so as to understand the teaching objectives and expectations of each schooling stage in the curriculum and master the ten key terms and corresponding mathematical thinking methods. The purpose is to accumulate sufficient theoretical knowledge for mathematics, so that when you are teaching a particular topic, you will be able to immediately relate the content to certain kinds of mathematical thinking methods.

(MasterMT, PD-trajectories excerpt, Ms C_{512})

Using the lesson preparation and implementation as carriers, I study carefully the curriculum standards, textbooks and the students. Each lesson is elaborately designed. Though I have been using the textbooks for many times, I still do an extra preparation before each lesson according to the characteristics of students, textbook content and the expectation of the curriculum standards. I also persist on making courseware and improving my professional competence through online learning.

(MasterMT, PD-trajectories excerpt, Ms Z_{105})

And indeed, studying the textbooks always serves the purpose of better preparation of teaching for learning:

In order to prepare a lesson, I always give relevant materials a sufficient reading and specify my lesson plan to sentences, taking into consideration all possible situations and corresponding measures . . .. I like the quote by Sukhomlynsky: 'It takes a lifetime to prepare a lesson.'

(MasterMT, PD-trajectories excerpt, Ms C_{310})

Ever since the commencement of my career, I have been teaching mathematics to the senior grades. To make sure every lesson is well taught, I give meticulous attention to lesson preparation every day and devote lots of effort into teaching research. When I first started, lesson planning is the most important thing of each day. To prepare for a lesson, I would read the textbook, the teachers' guide, and other references, before writing down the plan. This occupied pretty much all my spare time during the day and sometimes even the evening.

(MasterMT, PD-trajectories excerpt, Ms W$_{311}$)

It is important to carefully prepare every single lesson with great depth. In the lesson preparation, I not only study the students but also study the textbooks and teaching methods. Lesson types and teaching methods depend on the lesson content and actual status of students. I always write down carefully detailed procedures and timing of the teaching process, making sure the lesson plan is well written. Every lesson must be thoroughly prepared, with fun teaching tools being made to draw students' attention. I also stick to the habit of writing a summary after each lesson, making sure the key points and difficulty points are accurately dealt with. To design teaching methods and styles that fit the cognitive law of students, I pay much attention to reading various kinds of journals. I also seek to ease the difficulty points (难点) and emphasise the key points (重点). Carefully written teaching plans are followed by continuous summaries, which helps me improve teaching consistently.

(MasterMT, PD-trajectories excerpt, Ms L$_{210}$)

The better subject matter knowledge and pedagogical content knowledge of master teachers may be partially rooted in their consistent effort given to studying the curriculum and textbooks. With the enormous attention given to studying teaching materials as the overarching climate in China (Fan et al., 2015; Li, 2002, 2004), it is those who have studied deeper and better that become mathematically and pedagogically stronger than others.

Teachers perceive that their professional selves have shifted from experience-oriented to theory-oriented, as they move onto more advanced stage of professional development. Such transformation, according to them, is largely due to practice-based research. Their research, mostly qualitative, is different from what academics in higher education are doing. However, this form of research allows them to gain deeper understanding about their own practice.

9.8 Researching as practitioners: a different kind of research

Many teachers reflect on their experiences in doing research on their teaching practices often as part of a teaching research team. The kind of research teachers in China conduct is qualitative in nature, often in the form of case studies, with a focus on reflection and improvement over time. There are particular teaching research grants for teachers to apply for at district, municipal, provincial, and even national levels. Like other research grants in China, teaching research grant applications are put in by a PI who is supposed to name a team and their roles in the proposal. Often the projects aim at developing exemplary lessons and are supposed to be assessed according to the activities the team have carried out and results achieved. The implementation process thus heavily involves the PI and the team working together. The research projects make the teaching research activities more focused and systematic but quintessentially similar in that they share a similar mission – improving specific aspects of teaching in practice.

> Up until now, I have acted as a PI for three provincial projects and two municipal projects and worked as a major member on two projects.
>
> *(MasterMT, PD-trajectories excerpt, Ms C_{310})*

> Having received the advice and guidance from the experts, professors and renowned teachers, I started to do research. At the beginning, I took training courses; later, I started to undertake practice-based projects; now I am already leading my own projects as a PI. My effort is geared towards becoming a thinking, researching, practising and reflecting teacher.
>
> *(MasterMT, PD-trajectories excerpt, Ms S_{203})*

> As a member of the province's Central Group for Primary Mathematics, I have the opportunity to be a part of a number of national level projects. This has strengthened our team spirit and research competence. At the same time, I have successfully bid for and completed three provincial projects.
>
> *(MasterMT, PD-trajectories excerpt, Ms Q_{501})*

Practitioners as researchers, as in the case of action research, often risk bias and over-subjectivity, putting research findings down to a low level of generalisability. However, practice-based research by teachers themselves enables them to keep a critical eye on their practice and allows them to get to the root of the problem awaiting a solution. Given the depth of reflection and quality of refinement in practice, their findings have the potential to feed back to their own classes and be generalised to their local peers' classrooms.

9.9   Keep practising, keep changing

As teachers develop, professional awards, recognition and reputation flood in from the external world, somehow making continuous development harder for them than for a new teacher. Looking back at the PD process, master teachers wrote about the internal struggles they had been through and the level of perseverance that they had managed to reach ultimately. These teachers manage to constantly reflect upon and continuously scrutinise their teaching in greater details, seeking every opportunity to improve teaching at both micro and macro levels.

> To develop, a teacher must have time, space and the determination to learn. No matter how busy I am in work and life, I never hold up learning and

improvement. All the materials were accumulated using all the tiny moments squeezed in life.

(MasterMT, PD-trajectories excerpt, Ms Q₂₀₆)

Having become a subject leader, deputy head of school and senior teachers, many might think, there is no need to continue the hard work. For a very long period of time, I had lost motivation and stopped 'running', whilst at the same time a sense of unease arose, getting stronger and stronger. I found myself still posing questions to my class, still lacking confidence when facing the students. In the years to come, what should I rely upon to meet the challenges in primary mathematics education, the life-long career for me? As the text *The Eaglet Learning to Fly* says, 'Flying higher than the tree is not truly being able to fly.' I think I am not truly able to fly, so I need to think again. Since 2018, I have joined various PD programmes and seen these as opportunities to continue my queries into teaching practice and research.

(MasterMT, PD-trajectories excerpt, Ms C₃₁₀)

It is through diligent and continuous work that one can feel confident to go far. Einstein says the difference between human beings lies in their spare time. I believe that if a person spends the same amount of time each day in doing the same work throughout the life, this person is deemed to accomplish great achievements.

(MasterMT, PD-trajectories excerpt, Ms W₃₁₁)

It occurs to these teachers that there is always a better way of teaching. On the same content that is supposed to be delivered in 40 minutes, they would nit-pick all the smaller details for improvement on the ground of meeting learners' needs. They seem to be NEVER satisfied with their own practice. To them, teaching is a changing and therefore improvable practice instead of static and fixed status. Thus, they keep practising in order to stay afloat on the masterly 'ocean'.

9.10 Strong commitment to learner development and persistent interest in teaching

There is a strong sense of deep and persistent interest in teaching, at the beginning of each teacher's account of her/his path towards a master teacher.

I care about every student, and try to be the students' best friend.

(MasterMT, PD-trajectories excerpt, Ms Z₁₀₅)

The kind of love that a teacher gives to students must be not only strict but also respect, understanding and tolerance. My belief as a teacher is to help every child grow.

(MasterMT, PD-trajectories excerpt, Mr S₂₀₁)

In the class, I pay particular attention to motivating the students and communicating with them. Each lesson must fully reflect that learning is easily done by the children with fun and delight. I try to cultivate the competences of speaking and thinking as well as hands-on capabilities. To ensure every lesson is of good quality, I strive to develop children's comprehensive competencies and try my best to cultivate coherent and logical thinking amongst them.

(*MasterMT, PD-trajectories excerpt, Ms L_{210}*)

The most important thing to do as a teacher is keep the beginner's thoughts and dream – never forget why I chose the profession. This sheds a sacred light on the profession, sparking my heart.

(*MasterMT, PD-trajectories excerpt, Ms C_{310}*)

With such strong commitment, teachers have carefully sown the developmental elements into their everyday classes, which is evidence across the lessons we have observed. What they share with us during the interview tells us that, in their mind, students are not only learners in need of mathematical development but also young people deserving all-round development from early on.

9.11 Leading the way as a way of continuing learning and growing

In the interview or in written accounts about their PD trajectories, many teachers mention that they are in a leading position, such as the head of the department, head of teaching affairs or director of the teaching research group.

As a part-time teaching research coach rooted in the countryside, deep down I know being a Backbone Teacher in primary maths means not only being my best self but also leading a team so as to actively contribute to the innovation and development of primary maths teaching in the rural region. To do so, I have been to various rural primary schools and given numerous demonstration lessons and lectures, answering questions and offering advice. Over the past few years, I have mentored more than ten young teachers based outside of our school.

(*MasterMT, PD-trajectories excerpt, Ms W_{311}*)

2013 was a meaningful year for me when I became the head of the school's new branch. Amidst the numerous admin duties, I insist on taking the mission of teaching mathematics to one class and at the same time exploring collectively with colleagues the truth behind innovation. As a mentor, I assist my disciples [*students* as in Kungfu, the Chinese way of addressing young teachers who are mentored by a master teacher] in taking part in various teaching competitions, guiding the young teachers to grow professionally.

(*MasterMT, PD-trajectories excerpt, Ms T_{416}*)

Overall, three things contribute hugely to my development: consolidating a strong foundation of basic skills; actively taking part in competitions; mentoring young teachers.

(MasterMT, PD-trajectories excerpt, Mr W$_{418}$)

Our school set up a Master Teacher Studio in my name. In addition to organising events for our studio members, I share with everyone my teaching experiences and thoughts on teaching.

(MasterMT, PD-trajectories excerpt, Ms W$_{501}$)

As I gradually grow as a teacher, I started to work as a mentor for young teachers, guiding them to design lessons and grow.

(MasterMT, PD-trajectories excerpt, Ms S$_{505}$)

Research has shown that master teachers play a leading role in their peers' PD often through the activities run by the Master Teacher Studios (Cravens & Wang, 2017; Li et al., 2011; Zhang et al., 2021). It is meaningful to hear that master teachers find themselves benefitting from it, too, and they manage to refresh their minds and continuously develop as a result of leading others.

9.12 Master Teacher Studios: a cross-school professional learning community

Though each province may have a different MTS policy, it is similar across provinces that MTS hosts are approved about every three years after which they will have their achievement reviewed against the proposals that they put in three years ago. If they want to continue their role as a host, they will need to put in another proposal and get it accepted by the committee.

Of all the teachers who responded to our Trajectories survey, seven also provided an introduction to the development of their Master Teacher Studios and the teaching research activities run by their MTSs over approximately two years – from January 2018 up until April 2020 when the survey was sent out. To find the top ten Chinese words that featured in the MTS documents, we ran a word frequency on this part of the data and ended up identifying nine words by meaning as the two words both meaning 'teacher' were treated as one term (Figure 9.2). These words include 'master teacher studios', 'mathematics', 'members', 'primary schools', 'teaching research events', 'research', 'teaching research' and 'teachers'.

Next, we look at the development, status and activities of MTSs through two MTS cases: Ms W$_{314}$ from Jiangsu province and Ms L$_{407}$ from Jiangxi province.

9.12.1 Master Teacher Studio hosted by Ms W$_{314}$

Based in Yixing, Jiangsu Province, Ms W$_{314}$'s MTS was established in August 2019. She is amongst the second series of MTS hosts appointed in her city where MTS

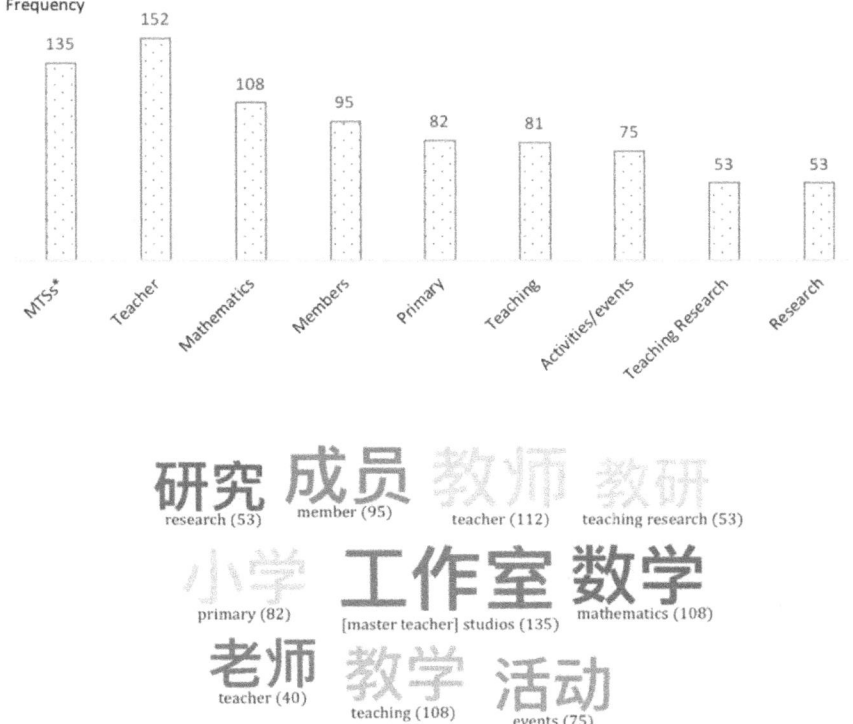

FIGURE 9.2   Words most frequently mentioned by teachers when reflecting on their MTSs

Note: MTSs = master teacher studios.

hosts are also regarded as supervisors to those joining the studios. The MTS has 14 members, with Ms W_{314} being the supervisor and the municipal teaching research official as a consultant. Of the members, three are already the Municipal Subject Leaders, five regarded as Rising Teaching Stars in the city, with the rest being young teachers with teaching experience of no more than three years.

Ms W_{314} cares deeply about rural education. With members mostly based in rural schools, her studio aims to cultivate category of teachers and the younger generation of teachers in the suburban areas of her city.

Every studio has particular missions that they state and justify in their studio proposals. In terms of learner development, Ms W_{314}'s studio sets their goal as developing higher-order thinking around the core knowledge of primary mathematics. In the ten events that her MTS hosted or took part in between August 2019 and April 2020, five events were about the development of higher-order thinking on key knowledge in mathematics classes. In November 2019, the entire team was in Guangxi Province, taking part in a nationwide teaching research conference on the teaching of calculations using the textbook series published by her home province (Jiangsu) – one of the major textbook series in the country. Two of the events took place in two different schools and consisted of demonstration lessons and expert lectures. Another event took place online, with the studio members sharing their recent readings.

Ms W_{314} is keen to lead the team to fly higher. Her writing is structured under three headings: (1) highly effective collaboration with clear responsibility and zero absence; (2) flying higher through collective effort, developing the knowledge and improving the practice; (3) mutual development whilst assisting teachers from underdeveloped schools and mutual improvement through collaboration and communication. In just one year, there are three teachers recognised as Rising Teaching Stars by the city and one teacher winning first place in the Provincial Excellent-Lesson Awards. They also published a lot: ten articles published by the professional journals in the province or beyond and over 20 papers winning awards from the city or beyond.

The studio has a three-year plan, studio rules and guidelines, including guidelines for meetings, for learning and exchange, for research projects, for attendance and promotion. Members are expected to be self-disciplined such that their work is effective and that '$1 + 1 > 2$'.

9.12.2 Master Teacher Studio hosted by Ms L_{407}

In her mid-30s, Ms L_{407} is already a Professoriate-Senior teacher and the deputy head (teaching) of her school based in Nanchang. She is diligent in doing research as a teacher. Her MTS was approved in 2015. The studio has expanded from the original six members including herself to now 20. She is keen in leading the team to do lesson studies. The team have been active in seeking opportunities to deliver demonstration lessons and take part in teaching competitions around the country. In addition, Ms L_{407} cares about the PD of rural teachers. She tries every opportunity to deliver demonstration lessons in rural schools, running teaching research activities with colleagues there. Her current team consist of 14 teachers from her own school, four from rural schools and two from other schools in the city.

To provide teachers, students and parents with open-access teaching and learning resources, she and her MTS members have learnt together how to record *micro-lessons* (微课). Then, the team filmed a series of micro-lesson videos summarising the textbook content of Grades 1–6. These videos can be particularly helpful for students on sick leave or in situations, such as the pandemic, where in-person teaching can be challenging.

Over the past two years, Ms L_{407}'s studio has carried out lesson studies focusing on the key lessons they identified: *Cumulative Law of Multiplication, Rotation of Shapes, The Clock and Time, Introduction to Cylinders* and *Introduction to Angles*.

All members take part in the meetings, studying and polishing lessons together. In addition, they have each observed other MTS members' lessons at least ten times, with some making it to 20 to 30 times.

They have observed, commented on and thereby polished the teaching research lessons they planned to deliver in rural schools as part of their voluntary support to underdeveloped schools. Each Tuesday during the two months before a new semester, the team organised textbook-analysis meetings, going through all the units in textbooks of the entire six grades. They arranged activities in supporting MTS members in submitting research proposals. The team have been to various teaching research events delivering research lessons, observing demonstration lessons and taking part in training sessions on excellent teaching and textbook use.

Moreover, Ms L_{407} has been voluntarily supporting her MTS members in polishing lessons for teaching competitions. As the head of the school responsible for teaching research, she also organised the regular schoolwide teaching research meetings/events, supporting colleagues in PD. In her already busy schedule, she has also accepted invitations and given lectures at teaching research conferences at municipal, provincial and national levels.

Across time, we have seen master teachers' PD trajectories and the development, status and activities of their MTSs as a cross-school PD community. Our final and next stop on the PD journey is three teaching research conferences at the municipal, provincial and national levels.

9.13 Participatory observation of teacher PD in action: teaching research conferences at various levels

Data on teaching research consist of four parts: (1) teaching research group (TRG) meetings at Ms Q's school (Chapter 2); (2) a municipal teaching research conference; (3) a provincial teaching research conference; (4) a national teaching research conference. With all events being observed through participatory observation, the former three were collected through in-person participation, whereas the final event online during the pandemic.

The typical professional learning community in China consists of TRGs or teaching research offices at the school, district, municipal and national levels. Teachers from the same subject group meet weekly or biweekly within schools. For a subject TRG in every school, there is usually a day in a week called the teaching research day which is essentially an afternoon on a school day, for example, each Wednesday afternoon. The agenda generally includes observing a teacher's lesson and making comments and/or revision plans. In the process of preparing for a teaching research conference or teaching competition at municipal, provincial or national levels, teachers will work together to assist one of them to plan and perfect a lesson. The school-based TRG will gather together to observe a number of versions of the lesson as the teacher teaches the same content in multiple classes. In urban schools, there are often more than six classes in a grade. So, if a school had six classes in

Grade 6, then the teacher would have six opportunities to deliver six versions of a lesson on the same content, say, *The Surface Area of a Column*. Besides school-based TRG meetings, there are teaching research conferences at district/municipal, provincial and national levels about twice a year.

Having been to two school-based TRG meetings in Ms Q's school, in the following sections, we will observe three teaching research conferences held at municipal, provincial and national levels.

9.14 Observing a municipal teaching research conference

In each semester, the municipal teaching research office will organise a teaching research conference. In Autumn 2018, the conference for the city happened to be hosted by our case study teacher's school. The conference takes place in the school hall on the morning of 21 September 2018.

The teaching research official, who took part in the TRG meeting in Ms Q's school (Chapter 2), gives an opening speech. According to him, last semester's teaching research conference focused on Geometry, whereas this semester's conference is about Numbers and Algebra. Three public lessons are presented, with each followed by a discussion session and expert comments from a panel consisting of the municipal teaching research official and renowned teachers from local. The first lesson is about addition and subtraction in Grade 3. The second and third lessons, both for fifth graders, are about *Representing Quantity with Letters* and *Solving Problems Involving Block Rates* respectively. Each lesson is delivered by a teacher from a different local school.

As in all publicly presented lessons, on the stage are the teacher and a class of students from the hosting school. The teacher and the students only get to know each other 5 to 10 minutes before the lesson. Students sit in rows as in the usual class facing one side of the stage where the board is placed (Figure 9.3a). On one side of the class, that is, the wall on the stage facing the audience, there is a large screen livestreaming what the camera from the other side of the stage captures – switching between a wide angle and a close shot of the teacher and students where necessary. Slides are also shown on the screen when necessary.

After each lesson, the teaching research official reappears on the stage with a microphone, asking if teachers in the audience would like to comment on the lesson. Teachers might be too shy to come forward, though everybody in the audience has been carefully taking notes as the professional convention here (Figure 9.3b).

The teaching research official thus gives a grand view over the location of the lesson content in the curriculum and asks why this lesson is necessary. He answers the question he posed regarding the difference between oral and written calculation for learning additions:

This specific lesson focuses on oral calculation of adding two-digit numbers. Students have learnt in Grade 2 to do the same type of tasks by hands. Why on

a

b

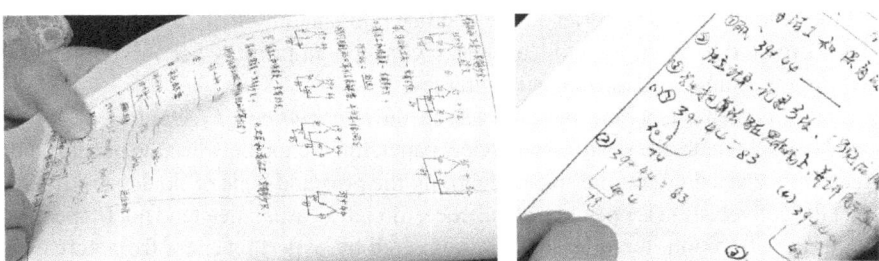

FIGURE 9.3 Teacher delivering a demonstration lesson on the stage with a class of unknown students

earth is the same content brought out again in Grade 3 in the form of oral calculation? My view is to lay the foundation for the introduction of 3-digit number addition. The focus rests on skills. More specifically, the rationale behind the calculation is the fundamental knowledge. Another thing needs attention, too. Is calculation by hands, which was already taught and learnt, the same as oral calculation? If that is the case, does it mean that we are repeating ourselves. Apparently, the two types of calculation are different. The former is done directly with the fixed methods, whereas the latter demands clear explanation and articulation in the process. Students are expected to not only do the calculation but also explain it.

Not much discussion results in the peer observation of each public lesson. Those who do speak up, for example the teaching research official, tend to offer insightful comments, since they are often former master teachers who know the nuts and bolts of the profession. The event seems to be more of broadening of teachers' vision with new ways of teaching 'old' content, generating thinking and small talk amongst them and those they are familiar with.

9.15 Observing a provincial teaching research conference

This is a 1.5-day provincial conference consisting of four public lessons, two invited lectures, three panel presentations by TRG leaders and teachers from three schools about their experiences of using a trial version of PEP's textbooks, and two presentations by a master teacher and the teaching research official about their award-winning research.

The conference opens with a panel of educational officials from local to the province sitting on the stage and greeting all teachers participating in person and online. The entire conference is being livestreamed such that it can be viewed simultaneously by all primary mathematics teachers across the province and literally anyone who has access to the Internet.

The first, main, part of the conference is a lecture by Mr W, the Editor-in-Chief of the leading textbook series in the country published by PEP. The lecture covers issues related to teaching and learning of concepts in primary school mathematics within the framework of numerical literacy.

Then comes a teaching research lesson on *Recognition of Cylinders*, following which Mr W chairs the post-lesson review panel. Interaction between the panel and the teacher is carried out on the stage, regarding the pros and cons of the lesson. It looks that the conversation is open to the audience, though as usual they tend not to speak up.

In the afternoon, three lessons are delivered by expert teachers from across the province, focusing on *Mathematical Thinking*, *Observation of 3D Shapes*, and *Definition and Meaning of Ratio* respectively. Following the public lessons are the 'lesson-sharing' presentations by two deputy headteachers from city N. Their presentations feature the lessons learnt in the course of implementing the trial version of the new textbook series to be published by the PEP. Then, the teaching research official from city J gives a presentation on the history of textbooks in the country over the period of almost 100 years. She and her team have done in-depth analyses of all the textbooks published during the period.

After the lesson sharing, Mr Wang, the provincial teaching research official and municipal teaching research officials from across the province sit on the stage as a panel, commenting on the three lessons delivered in the afternoon. This discussion concludes the first day's events.

The next day, a master teacher shares findings of her recent teaching-research project. The first author of this book delivers a lecture on her PhD study on the effectiveness of mathematics teaching in primary schools in China and England. Upon concluding the conference, the provincial teaching research official gives a presentation sharing his recent research. The final event sees the teaching research official formally handing out certificates of experimental schools to the second cohort of schools, which functions as a closing ceremony.

This one-and-a-half-day conference is packed with information and ideas for improving teaching. It offers time and space for teachers (both online and in person), textbook writers, educational officials and researchers to meet, discuss and debate on key issues in primary mathematics teaching and learning. The observable discussion and debate is however gentle and within a limited number of people, given the characteristics of Confucius culture and the power distance between professionals and their learners, and similarly between teachers and their leaders (Zheng et al., 2020).

9.16   Observing a national teaching research conference online

We have observed two national level teaching research conferences, with one hosted by a publisher in Nanchang in 2018 long before the pandemic and the other organised online by the National Committee for Primary Mathematics Teaching and Learning affiliated to the Chinese Educational Association. In a later stage of the project, the pandemic pushes various teaching research conferences online through livestream, allowing playback for a few days afterwards.

Here we present the second event as an example of national teaching research conferences. This recurring conference is usually hosted in spring biannually. The 2020 conference was postponed due to the pandemic, in the hope that an in-person event might be possible at some point in the near future. As time went by, the Committee decided to join the trend and move the conference online. It was scheduled from 22 to 28 of March 2021. To make sure all the demonstration lessons could be smoothly delivered, the conference organising committee asked provincial teaching research officials from across the country to send over the recorded lessons by teachers that were recommended.

Now that the first day of the conference has come, the committee starts to present lessons on their webpage in parallel sessions (Figure 9.4). The conference starts with a three-minute opening speech by two major figures in primary mathematics teaching in the country: Professor M from a well-regarded normal university in China and Ms W, the iconic master teacher who now chairs the National Primary Mathematics Teaching Committee.

Then, Ms W gives a public lecture titled Changing Learning Methods, Facilitating Deeper Learning. After pointing out the topic of the lecture, she begins with a short video showing a baby throwing pillows onto the floor multiple times in order to get off the bed safely with pillows as 'stairs'. This is used as a metaphor as offering children opportunity to learn for themselves by trying out their own methods and learning from their own experiences. The remainder of the lecture is theoretically presented around the thesis: placing children in the centre of education and offering them deeper learning opportunities (integration within mathematics) and wider application experiences (integration beyond mathematics). Detailed curricular examples are given and explained in the light of her thesis, making the lecture closely related to everyday teaching practice.

The slot following the lecture features the parallel virtual exhibition of three lessons on *Two-Digit by Two-Digit Multiplication*, *Introduction to Cuboids* and *Line*

Broadcast room	Content	Speaker	Provinces
Main Hall	Expert lecture: Changing Learning Methods, Facilitating Deeper Learning	Ms W	
	Two-Digit by Two-Digit Multiplication	T01	HLJ
	Introduction to Cuboids	T02	GD
	Line Graphs	T03	HB
Numbers & Algebra 1	Getting to Know Numbers up to Thousands	T04	JS
	Introduction to Decimals	T05	TJ
	Definition of Decimals	T06	SD
Numbers & Algebra 2	Preliminary Introduction to Fractions	T07	SX
	Reintroducing Fractions	T08	S'X
	Percentages	T09	FJ
Numbers & Algebra 3	Two-Digit by One-Digit Multiplication: Written Methods	T10	JL
	Multidigit by One-Digit Multiplication: Written Methods	T11	XJ
	Two-Digit by Two-Digit Multiplication: Written Methods	T12	GD
Numbers & Algebra 4	Introduction to Seconds	T13	GS
	Special Division	T14	SC
	Representing Numbers with Letters	T15	CQ
Shapes & Geometry 1	Introduction to Rectangles and Squares	T16	YN
	Finding the Shortest Perimeter	T17	AH
	Perimeter of Circle	T18	NM
Shapes & Geometry 2	Introduction to Acres	T19	GX
	Area of Triangle	T20	SD
	Columns and Cones (Review)	T21	HN
Shapes & Geometry 3	Axial Symmetry	T22	ZJ
	Movement of Shapes	T23	H'N
	Define Locations with Ordered Pairs	T24	H'B
Statistics & Probability	Introduction to the Mean	T25	BJ
	Reintroduction to the Mean	T26	LN
	Single Line Graphs	T27	H"N
Mathematical Thinking 1	Solving Problems Using Sets	T28	GZ
	Magic Möbius Band	T29	JX
	Number of Possible Situations	T30	SH
Mathematical Thinking 2	Chicken and Rabbits in One Cage	T31	QH
	Mathematical Thinking: Connecting Points into Segments	T32	XJ'
	Numbers and Shapes	T33	NX

FIGURE 9.4 Timetable of the 2021 national teaching research conference for primary mathematics (virtual)

Graphs respectively. In addition to the three lessons, there are 12 lessons about Numbers and Algebra, nine about Shapes and Geometry, three about Statistics and Probability, and six on Mathematical Thinking. These make a total of 33 lessons from 30 provinces or municipalities demonstrated and listed on the conference page, allowing for playback for a week.

9.17   It takes a village to nurture a master teacher

As our time-travel into the master teachers' PD trajectories and mathematics teachers' PD in action comes to an end, we realise that it is not just one element that contributes to better PD and masterly teaching. With continuous effort towards better teaching for better learning, master teachers are constantly resetting themselves to zero, always learning. All teachers channel their reflection through writing, researching and publishing. Intrinsically, it takes a teacher time, perseverance, diligence, practice, reflection, courage, and all the actions driven by these, to stay anew and masterful.

Extrinsically, it takes a village to make it happen. The tightly-connected teaching research community is readily available in every school for every subject and across administrative levels. The collective PD generates rich thoughts and catalyses self-reflection. The extrinsic conditions are mainly built through the close collaboration between teachers and their senior peers as well as the mentoring from expert/master teachers in the TRGs within schools and MTSs across schools. Support from school leaders and local teaching research officials who are former master teachers also play an irreplaceable role. Any benefits from teaching research activities are ultimately due to the *action* that it brings about. It functions due to the very action, just like teaching versus teachers in the making of good lessons (Stigler & Hiebert, 1999). The latter are able to make things happen only through the former. There are no other ways round. To give a demonstration lesson or prepare for a teaching competition, teachers have to polish a lesson up to countless versions with a supporting team often consisting of colleagues in the TRG/MTS, the TRG/MTS leader, teaching research officials at various levels depending on the level of competition. Polishing lessons offers the teachers opportunities to constantly reflect on what works better in greater detail and to receive feedback from abler, more experienced peers. Ultimately, teaching-research events at all levels focus and work exclusively on one thing – improving teaching and consequently learning of mathematics.

In the nationwide teaching research culture, the TRGs and MTSs are the real PD 'stations', for and by teachers, where they get their teaching competence constantly polished and, in a sense, fixed with assistance from peers and expert/master teachers. Teachers grow *collectively* in the security of these local stations, sharing the common purpose of practising the teaching craft better *independently* not only in their everyday classrooms but also in all kinds of novel settings, for example, in front of a panel of colleagues, experts, sometimes parents and perhaps everyone else off- and/or online in various teaching research events.

10

THE MASTERMT PROJECT AND THE FUTURE THAT'S ARRIVED

Chapter overview

- Master mathematics teachers: masterly mathematics teaching
- Master mathematics teachers: strong knowledge and beliefs
- Multiple learning outcomes: patterns and variation
- Masterly teaching, strong affective, metacognitive and cognitive outcomes
- Grow collectively, practise independently
- Destroying the teaching ego, together
- Keep reflecting, keep writing, keep improving
- Back to zero, moving forward
- One step forward: from effective to masterly teaching
- The future has arrived: Are we ready?

> The part of the grass above the earth basically grows rather slowly in the first half of its life when its root grows madly downwards to metres long. This is the part of its growth that is invisible to others. Halfway through its life when the raining season arrives, it is absolutely ready for upward growing, with its height reaching as high as two metres in days.
>
> (MasterMT, PD-trajectories excerpt, Ms C_{512})

Looking back at her professional development, Master Teacher, Ms C_{512}, describes her growth metaphorically as that of the elephant grass in Africa. It is a process of perfecting teaching through patiently accumulating knowledge and skills over an enormous period of time without drawing much attention or chasing fames and

DOI: 10.4324/9781003127925-10

recognition from the external world, as if she had stepped into the status of flow (Csikszentmihalyi, 1990) but at a grander scale – in years or, more precisely, decades. It is not about reaching a destination. The process itself is the destination. The essence to it is continuity.

As our journey into master teachers' practice and PD comes to an end, looking back, the findings of the project are substantial, with the breadth and depth that we could most possibly reach in one project. The project has robust statistical models, generated with MLM and MSEM. These models indicate causal relations between teaching and three types of learning outcomes, namely affective, metacognitive and cognitive outcomes in mathematics, with some interactions between these learning outcomes identified as well. Equally substantial but much richer findings resulted in our qualitative analyses of lesson videos, post-lesson interviews with teachers, teacher accounts of their PD trajectories towards masterly teaching – we received a total of 69,903 words of writing in Chinese from them on PD alone – and our participatory observation of 'PD in action' at the school, municipal, provincial and national levels before and during the pandemic.

As discussed earlier, there is a limited amount of work out there on Shanghai teachers' PD which acknowledges the leading role that master teachers play in their peers' PD. Albeit informative, this body of work currently leans towards loose policy reviews or qualitative research about several cases. Worldwide, we are the first to report large-scale findings on master teachers' teaching practice, knowledge, PD trajectories and their influence on colleagues' PD. Findings from existing studies on maths teaching were based on either quantitative (QUAN) or qualitative (QUAL) methods. These include the widely known TIMSS Video Studies (QUAN) (Hiebert et al., 2003; Stigler & Hiebert, 1999), the Learner's Perspective Study (QUAL) (Clarke et al., 2006), and the recent TALIS video study (QUAN) (OECD, 2020). We have made the best use of both QUAN and QUAL methods, which makes the study much needed in methodological terms and both rich and robust in terms of findings, implications and contributions.

The holistic reality gradually emerges, as multiple measurements join each other through multiple levels and multiple perspectives regarding maths teaching join each other across multiple layers. The beginning joins the end, with the holistic reality of masterly teaching represented by the integrated research findings:

- master mathematics teachers are indeed teaching masterfully in their everyday classrooms;
- master teachers have strong knowledge about mathematics as a system, about learners and learning, about how to teach, with a genuine constructivist belief system at heart and a strong reflective-practitioner mindset;
- students from their classes are high performers in all three categories of learning outcomes that we measured in comparison with published results using the same instruments/items;

- multilevel models and (multilevel) structural equation models show that teaching does make a substantial difference in learners' affective, cognitive and meta-cognitive outcomes in mathematics;
- based on their PD trajectories and PD in action, one major secret to teaching in a masterly way is growing collectively and practising independently;
- another major secret to the master teachers' success is that they are brave enough to destroy the teaching ego together with their peers in order to rebuild and improve their teaching practice;
- they read, research and write about their practice, which makes them reflective practitioners who know where and how to improve;
- they constantly reset themselves to zero so that they can look at teaching anew and keep improving;
- one last major secret is that they see effective teaching as a threshold above which is their goal, masterly teaching.

In what follows, we will discuss these key findings sequentially and conclude upon the future that is already in master mathematics teachers' classes.

10.1   Master mathematics teachers: masterly mathematics teaching

In master teachers' classes, students have rich opportunities to learn, through whole-class (including the teacher) exploration into the mathematical world. The thinking-out-loud energy runs throughout the class, and students' attention is constantly drawn towards the mathematics that is under exploration and discussion, which keeps them firmly on task. The thorough discussion during the whole-class interaction deepens and sharpens their thinking and gets them ready for independent work that is completed in almost an instant due to profound understanding. The group work takes place at the right moment when all possible answers, either right or wrong, are likely to emerge. The emerging answers then work as teaching and learning resources in the subsequent whole-class discussion. The rich collection of answers both widens and deepens students' understanding, as various answers join each other in the whole-class 'peer-reviewing' process.

Master teachers demonstrate a strong competence in formatively assessing learning as the lesson unfolds. Lessons are clearly structured, presented and articulated, with optimally sequenced tasks and activities. The class discourse offers rich opportunities for students to clarify, justify and synthesise ideas, which enhances the strengths and clarity of the thinking mind. The clarity of class discourse is mostly driven by the chains of teacher questioning which engages students in coherent thinking around the essence of knowledge embedded in the focal task(s). Teachers demonstrate strong teaching skills to keep the collective 'thinking-ball' rolling onwards. Their teaching skills are 'disguised' in the forms of process questions they ask, the tasks they give at the optimal moments in time, and the presentation of and discussion around student work, again at the optimal moments in time. Everything seems to naturally unfold, which is, however, all deeply rooted in the strong teaching skills that are apparently refined over time. In responding to questions from the teacher and sometimes from peers, students are constantly thinking and talking about their thinking and then monitoring and regulating their cognition once they have a better understanding of their own and their peers' ideas about a particular category of mathematics problems and the underlying knowledge. Students are expected to work at their best, and they voluntarily contribute to the collective discourse on the focal topics. Such collective concentration leads to a positive learning climate in the class, which makes it unnecessary for the teacher to have to maintain discipline in the class.

Master teachers appear to build an optimal connection between students and mathematics, offering quality mathematics teaching. Their lessons show rich representations of mathematics, connections within mathematics, and connections between mathematics and the real world. As the students think along and respond to teachers' questions and as they present their work to the class with thoughtful explanation and justification, the mathematical contents are unfolded in the class in a clear and accurate manner. Learning difficulties are identified and collectively resolved over the course of students presenting their work in the class and reviewing errors that emerge in the work presented. Students are thus able to learn from both their own and their peers' successes and failures in the reflection stage of problem solving. With thorough understanding of the mathematics that underpins a series of problems, they are thus able to pose problems that serve to check the robustness of the conceptual understanding by embedding some of the 'pitfalls' that they or their peers have previously fallen into. These problems often expect accurate, coherent and complex reasoning.

Prioritising the key points of learning, master teachers seek to model the way for a shared discourse, by clarifying beforehand work expectations or providing necessary communication symbols/terms that are similar to what Ausubel (1963) called advanced organisers. Through class presentation and discussion of multiple representations, students gain deeper understanding of the essential mathematics underpinning the representations. In a class full of loops of questioning-response-feedback interactions, students develop fast in both cognitive and metacognitive terms. Teachers organise the mathematical tasks in an optimal order, such that students naturally move on to the cognitive milestones without over-struggling. The tasks are organised in an optimal sequence which seems unlikely if without thoughtful planning beforehand. All in all, teachers demonstrate a strong competence in diagnosing the cognitive status of their students (Leuders et al., 2018).

In the master teachers' classes, students do not move on until they have reasoned in-depth the very essence of the task/topic. At the core of the lesson lies the key structure of mathematics, rather than fragmented pieces of knowledge. Despite the simultaneous use of smartboards and projectors, teachers still make use of the chalkboard to gradually unfold the essence of knowledge. In the revision lessons, master teachers formulate the learning experience such that the class revisit the learned knowledge from abstract to concrete, from theory to practice, from deductive reasoning and inductive reasoning, and vice versa. Across classes, it is remarkably important for the students to know and understand not only the *what* but also the *how* and *why*.

Knowing the *what* is thus realised through understanding, knowing the *how* through understanding-based practice, and knowing the *why* through reasoning, hypothesising, hypothesis-testing and debating. Instantaneous learning that emerges in the lesson is immediately turned into learning and teaching resources. Another feature of revision lessons is that students are able to construct a personalised knowledge map with greater accuracy and strong logic and to pose high-quality mathematics problems. The lessons introducing new knowledge progress by offering students rich learning experiences, from hands-on to heads-on and from written maths to mental maths, such that the knowledge can be ultimately internalised by the learners after thorough understanding. In the lessons aiming for practice of newly learnt knowledge or skills, practice is done through enriched reasoning about the rationale behind the knowledge or skills, leaving no room for just mechanical practice. Every step of practice is questioned and justified by the teacher and students. All classes featured the constant use of peer review of student work during the whole-class discussion time following the independent or group work.

10.2 Master mathematics teachers: strong knowledge and beliefs

Master teachers have a rich system of knowledge. They seem to know well not only mathematics (both in the abstract sense and in the real world) but also children. And they seem to have channelled an enormous chunk of their thoughts towards their bridging role between the two. During the interviews, all teachers talk about mathematics, children and connections within and between them. As teachers articulate their thoughts on teaching, they make strong pedagogical reasoning (Ball, 1988). As they reflect on what students have learnt previously, are about to learn 'now' and will be learning in the future, they make strong metacognitive reasoning, about others', in this case students', cognition. It is clear that they have been planning and providing conditions for a certain kind of and a specific level of cognition – thinking, reasoning or understanding – to happen 'now' and develop into the future.

It is also apparent teachers themselves have put enormous effort into the learning of the subject both as a discipline and as compiled in the school curriculum and textbooks. During the planning, implementing and reflecting stages of the lesson, they are free to retrieve information in the internalised and connected system of mathematical knowledge. They see children's learning as a dynamic and longitudinal process and work hard to diagnose their status and 'history' of learning such that their teaching builds best connections between children and mathematics.

In addition to absorbing the system of knowledge in mind, they have a system of goals to accomplish through teaching. Mathematics must be learnt well – not only the basics but also the higher-order; not only the abstract but also the application;

not only solving challenging problems but also identifying and posing high-quality problems; not only as knowledge points but also as a connected system of knowledge; not only knowing the *what* but also the *how* and *why*. Their rationale for their teaching shows strong evidence for strong metacognitive development amongst students. Learners in their classes are constantly promoted to learn to learn strategically and to learn from not only one's own but also others' experiences. After all, they hope their students can learn something from the mathematics class – something enjoyable and meaningful to themselves and something useful in their life not only now but also in future. They hope the children learn to respect others in the process of learning in the maths class. They seem to be bearing the mission to sow the moral, happy, learning and reflective seeds in children's minds.

Master teachers are both *reflective teachers and learners*. Across the interviews with the teachers, we have a common impression that they embody the two roles in themselves. On the one hand, they are reflective teachers who keep thinking and reflecting throughout the lesson and even more so before and afterwards. On the other, they seem to see themselves as humble learners who they assume always have gaps in knowledge and skills, therefore constantly in need of learning. This makes them reflective learners who look deeply into learners and learning as well as themselves and teaching. Because of the abundant question-driven interaction and rich opportunities to discuss each other's work, their students are *reflective learners and teachers* to each other. They have developed into reflective learners, who not only attend to the *what* but also the *how* and the *why* questions thoughtfully, as well as the reflective teachers who develop, explain and justify mathematics to other learners.

10.3   Multiple learning outcomes: patterns and variation

In Chapters 5–7, we have measured three categories of learning outcomes – affective, metacognitive and cognitive learning outcomes – and identified major patterns and variations.

The average rating (2.5, $SD = 0.6$) of attitudes towards mathematics is quite high on a scale of 3. In terms of metacognition, the students reached a much higher score on each of the 18 Jr. MAI items than those from a culturally different society, such as the United States (Sperling et al., 2002), and those from a culturally similar society, such as Singapore (Ning, 2016) where the same construct was developed and utilised. It suggests that students taught by master mathematics teachers are indeed metacognitively more competent in mathematics learning than those otherwise. They are well aware of their cognition and able to regulate their thinking in the learning processes. However, there is, as we anticipated, variation by gender, grade and SES. In mathematics, students from master mathematics teachers' classes demonstrate a strong level of mastery. Their correct rates on each of the tested items were much higher than the correct rates of their senior peers in British schools who were two to five years older than them at two points in time (1976 and 2009).

There are variations in all three categories of learning outcomes primarily affected by student background factors, that is, age, gender or SES. Apart from the tiny effect of age on perceived teaching engagement $(r = 0.04, p < 0.05)$, there are little effects of it on either school belonging or attitudes towards mathematics. In line with previous research, the cross-sectional data suggest that age plays a significant role in learning and that children do better metacognitively $(r = 0.10, p < 0.01)$ and cognitively $(r = 0.33, p < 0.01)$ in mathematics as they grow older. In mathematics learning, girls outperform boys metacognitively (3.99 vs. 3.91 out of 4, $t(3022) = 3.42, p < 0.001, d = 0.13$) whereas boys scored higher on attitudes towards mathematics (2.57 vs. 2.49 out of 3, $t(2777) = 3.90$, $p < 0.01$, $d = 0.15$) and do better cognitively in Grade 5 (467 vs. 449, $t(1306) = 64.58, p < 0.001, d = 0.27$) but similar to girls in Grade 6 (545 vs. 539, $t(1320) = 1.03, p = 0.30$). In terms of perceived schooling and teaching on the scale of 3, girls tend to have a higher sense of school belonging (2.69 vs. 2.59, $t(2934) = 6.21, p < 0.001, d = 0.23$) and feel more engaged by the way their maths teachers teach (2.67 vs. 2.62, $t(2589) = 2.94$, $p < 0.01$, $d = 0.11$) than boys. Student perceptions of schooling, teaching and mathematics learning show no connection with their SES. However, their cognitive $(r = 0.23, p < 0.01)$ and metacognitive $(r = 0.19, p < 0.01)$ outcomes are significantly affected by socioeconomic background.

In comparison with previous research, the master teachers' students demonstrate relatively high performance in the three kinds of mathematics learning. There are, however, still variations of outcomes by age, gender and SES to a varying degree. Keeping the patterns and variations of learning outcomes in mind, we have sought to examine the effects of teaching on learning through multilevel modelling and (multilevel) structural equation modelling, in addition to qualitative interpretation of the data.

10.4 Masterly teaching, strong affective, metacognitive and cognitive outcomes

After controlling for student age, gender and SES in Chapter 5, the collective perception of teaching engagement explains 96.7% of variation in student attitudes towards mathematics at the class level. This means even amongst the master mathematics teachers there is still considerable between-class variation in the cultivation of on affective learning outcomes and hence room for improvement for some of the teachers in this regard. The perceived teaching engagement also plays a mediating role between student sense of school belonging and their attitudes towards mathematics learning.

In Chapter 6, we confirmed that students with strong performance in metacognition have better understanding and regulation of their own thinking, planning and learning (Flavell, 1976, 1979; Sperling et al., 2002). Master maths teachers' students generally demonstrated quite high self-reported features of metacognition, having good knowledge and regulation of their own cognition during the process of mathematics learning. Metacognitive outcomes varied by age, socioeconomic status and gender, with girls doing better than boys. Our analyses further confirmed that with metacognition a meaningful distinction can be made between knowledge and regulation of cognition. Furthermore, teaching engagement explained a sizeable portion (40%) of metacognitive performance at the class level, controlling for age, gender and SES, with the association being *indirect*: metacognition-oriented teaching predicted student-perceived teaching engagement which in turn predicted metacognitive learning outcomes. The findings are in line with previous research that emphasises the importance of metacognition in the mathematics classroom (Desoete et al., 2019; Muijs & Bokhove, 2020; Zhao et al., 2019). One thing we can take away from this is that the primary mathematics classroom in China integrates metacognitive strategies and is a place where metacognition has a firm place in the instructional strategies. In adopting the qualities of such classrooms, teachers should be aware of and have such metacognitive skills, as to explicitly implement them in the classroom.

The multilevel modelling in Chapter 7 indicates that 29% of variation in cognitive learning outcomes is explained by the teacher level variables. Controlling for student age, gender and SES, the quality of teaching or mathematical teaching further explains 16% to 17% of mathematics performance. The results of multilevel structural equation modelling further unveil the indirect effect of teaching on mathematics performance mediated by the effect of teaching on metacognition.

The integration of quantitative and qualitative findings indicates that master maths teachers take the teaching-learning mechanism to a whole new level where they pretend to be learners and learn with the students/learners who have already been used to taking up some of the teaching role, presenting, explaining, asking each other questions, debating and concluding on mathematics. This mechanism, apparently formulated over time, puts children firmly at the centre of the class but benefits from covert yet carefully planned guidance from the teacher. When the teacher 'hides' her/his teaching (or scaffolding) under the 'disguise' of questioning and inviting children to answer and present, the entire class perceive the underlying lesson goals as a collective mission, working at their best in leading the thinking and realising the mission.

10.5 Grow collectively, practise independently

Findings of the MasterMT project indicate that master mathematics teachers have indeed mastered mathematics teaching and are constantly pushing for improvement. Their development into master teachers is a combined result of internal and external forces.

All master mathematics teachers attribute their growth partly to the external support from their senior colleagues, peers and teaching research leaders. However, it would not be possible for them to become master teachers, if without the constant drive for professional excellence from within and the actions they have taken in the pursuit of excellence.

Within and across schools, the TRG- or MTS-based PD offers teachers sustainable opportunities to improve teaching collectively. Their improvement is content specific, which involves an intensive study of curriculum, textbooks, learners and learning over time, regarding the teaching of a specific topic, say, *Adding and Subtracting Fractions with Unlike Denominators*.

Existing reviews of empirical studies on teacher PD present a mixed and somewhat contradictory picture, with PD being identified either as a contributor to student achievement (e.g., Yoon et al., 2007) or not necessarily so (e.g., Sims & Fletcher-Wood, 2021). When focusing on instructional standards, teacher-led collaborative enquiry with peers was found to be a value-added predictor of teacher effectiveness (Cravens & Hunter, 2021). According to Sims and Fletcher-Wood (2021), repeated practice might work better than sustained PD over a long period of time. However, repeated practice could be problematic when poorly defined. If it means perseverance in perfecting one's practice in a focal area, then that seems to coincide with what we found in our study. If it means literally 'repeating' practice without thoughtful plans for improvement, then it is far from the kind of PD that master mathematics teachers and their peers are doing.

In *school-based PD* events we observed, it is about developing a newer and – from the practitioners' points of view – better version of the focal lesson. What teachers have been doing is seeking opportunities to polish one particular lesson for, indeed, numerous times in order to develop an exemplary lesson that can be presented to a panel of experts and a wider audience in teaching research events or teaching competitions. The process involves cycles of designing-implementing-reflecting-revising-implementing-reflecting that take weeks, months, and even semesters, but it is never simple repeats of what has been done. Each version of the lesson is built on the former version(s) from which they always identify crucial points of improvement. Thus, each version is different from, rather than repeating, what was done.

Teaching research conferences expand teachers' vision as to how innovatively a lesson can be designed and carried out. Nevertheless, it is certainly the school-based TRGs and cross-school MTSs that make the major contribution to the cultivation of masterly teaching and master teachers, in addition to teachers' self-driven effort. These professional learning communities create concrete opportunities for teachers to make genuine refinement and continuous development in teaching, generating invaluable advice from peers and crucially from established master teachers. It is by locating PD programmes naturally and longitudinally in or close to the original habitats (i.e., schools) of teachers that real teaching improvement and professional development can happen and sustain the test of time.

10.6 Destroying the teaching ego, together

Everything about the teaching research events is based on the shared value that teaching is a public activity – an activity that is relevant to the public and therefore not solely the teacher's own business. This is a fact that generates no argument, since children are the future generation of the public and their education is a matter of collective interest of the public.

With the shared value comes the collective sense that together the profession will grow as every member of it is putting their best effort into *teaching*. It takes a strong team to polish a public or teaching research lesson. First, a team, consisting of the teacher, experienced teaching research officials and master teachers, prepares a demonstration lesson carefully and revises it probably at least 10–20 times before reaching the final version. Then, the lessons by different teachers are sent to the reviewing committee who will select the best collection of lessons that will be shown to the public – by public they mean everyone in the maths teaching profession and education. The scope of public reach can be the wider society, since anyone who has access to the Internet can access and see the livestream or playback at a preferred location.

Because of the intensive preparation behind every demonstration, each lesson is an ideal version showcasing new ideas for teaching a particular topic, like a concept car in the automotive industry. Albeit a version that may be hard to achieve in everyday teaching, it opens up the professional minds and allows teachers to see the room for and the possibility of improvement regarding their everyday practice. Others may simply copy the version into their daily teaching. It is okay to 'copycat' teaching ideas from others, as long as it is not in a public demonstration that is supposed to be original and stamped with a personal mark. In this profession, when talking about the teaching of a certain topic, say *Fraction Division*, teachers often refer to teaching in a specific way by the teacher's name, for example, Teacher Wang's version.

In front of an audience of fellow teachers, the teaching-research lessons are delivered by excellent teachers nominated by schools and districts. The audience could range from about 20 mathematics teachers in the same school if it's a school-wide event to a combination of hundreds of teachers in a large school hall and tens of thousands of other colleagues online if it's a provincial or national event.

Teachers are supposed to teach children often from the 'host' school whom they are not supposed to have taught before and whom they are only allowed to meet and get familiar with 5–10 minutes before the actual lesson on stage. It's all about removing any settings that the teacher is familiar with and letting the teacher prove her/his teaching competence with 'brand new' students in a novel environment and the public's eyes.

The underlying assumption seems to be that the teaching masters should be able to adapt to and survive any conditions, without sacrificing the quality of teaching. To them, teaching is a live and adaptive system. With solid knowledge and skills, they are ready to lift teaching from Science to Art! In this sense, if we are going to cease the teaching-as-science-or-art debate, their solution is to treat 'teaching as science' as a threshold above which is the free space for 'teaching as art'.

Anywhere, with any students of the hypothesised age, the teacher should be able to teach in an excellent way. Not just effective, but brilliant, outstanding, new and creative! Effectiveness is simply the baseline of the mission. This is probably why originality plays an essential role – because everyone in the audience gets a chance to pick the wrong bits, few would risk copying when everything is under the spotlight on the stage and livestreaming 'worldwide' and when it's really an opportunity to be known by everyone else in the profession. All recognised lessons must be fantastic in their own way! After the lessons, teachers are supposed to communicate with experts on stage and volunteers from the audience, answering questions and sometimes justifying the design and implementation of the lesson.

Teaching research (教研), or more precisely translated, 'teaching study', is a tradition that goes deep into every cell of Chinese schools. And it keeps evolving with technology and time. It looks at the everyday practice of mathematics teaching like the idea of home cooking – easy, comfy, hearty and brilliant in all possible ways, in its own way. Teaching-research-community-based PD is probably one of the most vibrant territories in China's education that is truly outstanding and cutting-edge. It is a territory that achieves its status by constantly pushing boundaries for and by the practitioners. It gets where it stands because, amongst other reasons, it is run by teachers and led by master teachers and teaching research officials who are supposed to be former master teachers such that the fellow teachers would naturally see them as role models, respect them and genuinely want to collaborate with them. With the relatively new addition of Master Teacher Studios, teachers have stronger support and richer opportunities to pursue professional development within and across schools.

10.7 Keep reflecting, keep writing, keep improving

Teacher-led professional research projects allow teachers to do a different kind of research from the kind conducted by researchers in academia. Their work is more subjective and practice-based, demanding more reflection by the practitioner as a researcher (Hammersley, 1993). Though their research outputs are less likely to be generalisable, their teaching practice is strengthened over time because of deeper reflection and thinking which has resulted in the self-researching process. Reflective practitioners are likely to understand and improve their practice better (Schön, 1983).

If demonstration lessons and lessons for teaching competitions push teachers to polish their practice, then practice-based research, reading, writing and publishing push them to polish their thinking and reasoning about the teaching materials, targets and processes more often and to a greater extent than otherwise. These together lead to continuous improvement of teaching.

Teachers certainly benefit from in-depth study into their own practice, but one should be cautious about the generalisability of findings of such research. We shall not encourage teachers to learn to do the kind of research that academic researchers are doing when the ultimate goal is to improve their practice. With teaching improvement as a goal, teachers should be encouraged to research and reflect on their own teaching practices together with peers. Their refined lessons and thoughts offer thoughtful examples for their peers, once published and shared in the teaching community.

10.8   Back to zero, moving forward

> The only true wisdom is in knowing you know nothing.
>
> – Socrates

Master teachers are able to reset their mental status as a teacher and keep their mind open to new approaches to teaching and learning. This offers them a chance to review their own practice which has already attained various regional and national awards. As a result, their mind is constantly refreshed and ready for new adventures in the teaching-verse.

To them, *each lesson is different*. Teachers move from studying and mimicking expert teachers' teaching to building their own styles and then to debunking their static styles. This way they are able to evolve. They aim to develop a unique way of teaching each version of each lesson, seeing each lesson as a different case. As Ms W_{311} puts it, 'Each lesson is set in a unique context. No lesson can be replicated.' From an empirical research perspective, researchers, especially those inhabiting in the quantitative paradigm, would believe that there are things that can be generalised and replicable. However, for practitioners, it might be key to see each lesson uniquely designed for a specific class of students in their specific status in the specific period of time and specific space. This way, best teaching happens as the teacher attempts to tailor the ongoing activities as she/he sees fit, taking into account all the specific micro conditions. Perhaps, what the teachers might not see are those elements of their teaching that are constant. Even the attention to personalise each lesson on each occasion is in and of itself a constant behaviour. The sense of not replicating oneself like a machine might be important in any human activities.

10.9   One step forward: from effective to masterly teaching

Knowing how to teach mathematics well to children makes a teacher effective in teaching. Being authentic and creative makes the effective teacher a fun human being to learn with. Beyond the effectiveness threshold, this fun person is able to personalise specific learning experiences for a particular class of children in a specific time of a year, at a particular location. Beyond the science of teaching, one step forward is the art of teaching; without science as a basis, there is no freedom in the land of teaching as an art. Teaching masters, once apprentices, grow into outliers, rising from their practice (Gladwell, 2008). One similarity amongst many is that they all feel the constant urge to improve upon their former practices, to be better, constantly. Their standards on teaching fly far beyond effectiveness. The constant urge for perfection drives them towards masterly teaching that keeps evolving in spite of more and more recognition from peers and professional bodies.

10.10   The future has arrived: are we ready?

The futuristic mathematics teaching being dreamed about by researchers, policy-makers, parents and the like is already a reality in master teachers' classrooms. Many of the demands in the latest curriculum standards in various countries (e.g., DfE UK, 2013; MoE China, 2022; MoE Singapore, 2012; NGA & CCSSO, 2010) are already an established reality in master mathematics teachers' classes as observed in spring 2019. Those include but are not limited to mathematical modelling, self-regulation, quantitative competence, abstract reasoning, spatial reasoning and competence, mathematical application, and so forth.

The ideal constructivist learning is observed in action in master teachers' classes. In their classes, learning is naturally constructed by the learners with the scaffold provided by the teacher. This much longed-for feature is universally realised across these classrooms where knowledge is constructed by the learners. However, the teacher's role is more important in such learner-led constructivism than in classrooms without learner-led constructivism or with a lesser degree of it. The key lever is the teacher-designed scaffold. The scaffold is a complex system of teaching and learning material, consisting of, but again not limited to, lesson plans at macro, meso and micro levels, tasks and the dynamic adjustment of small steps. Once again, master teachers are able to avoid the classical either-or trap, in this case, influencing the learning process *versus* enabling learners to construct knowledge. In master teachers' classes, both happen, coexist and co-work, for one overarching purpose – learning.

The TRG- or MTS-based PD allows two crucial things, amongst many, to happen. First, it provides teachers opportunities to develop the best version of their teaching on each focal topic that is chosen. The word 'best' here is a similar concept to infinity in mathematics which one can constantly get closer to but can never actually arrive at. Hence, there is always room for improvement. This works on the individual teacher's level, such that every teacher gets better than 'yesterday'. Second, at the PLC level, each TRG or MTS feeds the best practice ingredients back to itself, with the leading figures, often master teachers, demonstrating the best teaching practice. Both approaches take much shorter turnover time than the rather strict randomised control trials that seem to be more likely accepted by academia-oriented teaching improvement. The master-teacher-led PD works because it effectively prevents teachers from 'always starting over again' (Cai et al., 2020) by constantly accumulating and refining the best practice recipes within and across generations of those in the profession.

With the latest evidence from master teachers' classes, we shall be confident to say that it is time to embrace the future, now. The next step is to generate the futuristic features at scale through close collaboration between researchers, teacher educators, parents, practitioners and policy makers, and most importantly between teachers, allowing best mathematics teaching practice to scale from the bottom up by and for teachers.

REFERENCES

Akyol, G., Sungur, S., & Tekkaya, C. (2010). The contribution of cognitive and metacognitive strategy use to students' science achievement. *Educational Research and Evaluation, 16*(1), 1–21.

Allen, K., Kern, M. L., Vella-Brodrick, D., Hattie, J., & Waters, L. (2018). What schools need to know about fostering school belonging: A meta-analysis. *Educational Psychology Review, 30*(1), 1–34.

Anderson, J. R., Betts, S., Ferris, J. L., & Fincham, J. M. (2011). Cognitive and metacognitive activity in mathematical problem solving: Prefrontal and parietal patterns. *Cognitive, Affective, & Behavioral Neuroscience, 11*(1), 52–67.

Askew, M. (2013). Big ideas in primary mathematics: Issues and directions. *Perspectives in Education, 31*(3), 5–18.

Askew, M., Rhodes, V., Brown, M., Wiliam, D., & Johnson, D. (1997). *Effective teachers of numeracy: Report of a study carried out for the teacher training agency.* London: King's College, University of London.

Ausubel, D. P. (1963). Cognitive structure and the facilitation of meaningful verbal learning. *Journal of Teacher Education, 14*(2), 217–222.

Ball, D. L. (1988). *Knowledge and reasoning in mathematical pedagogy: Examining what prospective teachers bring to teacher education* [Doctoral dissertation, Michigan State University].

Ball, D. L., Thames, M. H., & Phelps, G. (2008). Content knowledge for teaching: What makes it special? *Journal of Teacher Education, 59*(5), 389–407.

Bates, D., Mächler, M., Bolker, B., & Walker, S. (2015). Fitting linear mixed-effects models using lme4. *Journal of Statistical Software, 67*(1), 1–48. https://doi.org/10.18637/jss.v067.i01

Behr, M. J., Harel, G., Post, T., & Lesh, R. (1992). Rational number, ratio, and proportion. In D. A. Grouws (Ed.), *Handbook of research on mathematics teaching and learning: A project of the national council of teachers of mathematics* (pp. 296–333). Reston, VA: National Council of Teachers of Mathematics.

Bentler, P. M., & Bonett, D. G. (1980). Significance tests and goodness of fit in the analysis of covariance structures. *Psychological Bulletin, 88*(3), 588.

Berliner, D. C. (2001). Learning about and learning from expert teachers. *International Journal of Educational Research, 35*(5), 463–482.

Beswick, K. (2006). Changes in preservice teachers' attitudes and beliefs: The net impact of two mathematics education units and intervening experiences. *School Science and Mathematics, 106*(1), 36–47.

Beswick, K., & Fraser, S. (2019). Developing mathematics teachers' 21st century competence for teaching in STEM contexts. *ZDM, 51*(6), 955–965.

Biggs, J. B. (1982). *Evaluating the quality of learning: The SOLO Taxonomy*. New York, NY: Academic Press.

Bijlsma, H. J. E., Glas, C. A. W., & Visscher, A. J. (2022). Factors related to differences in digitally measured student perceptions of teaching quality. *School Effectiveness and School Improvement, 33*(3), 360–380.

Bond, L., Smith, T., Baker, W. K., & Hattie, J. A. (2000). *The certification system of the national board for professional teaching standards: A construct and consequential validity study*. Greensboro, NC: Greensboro Center for Educational Research and Evaluation, University of North Carolina at Greensboro.

Bond, T. G., & Fox, C. M. (2015). *Applying the Rasch model: Fundamental measurement in the human sciences*. London: Routledge.

Boyer, C. B. (1968). *A history of mathematics*. New York, NY: John Wiley & Sons.

Bunge, S. A., Wendelken, C., Badre, D., & Wagner, A. D. (2004). Analogical reasoning and prefrontal cortex: Evidence for separable retrieval and integration mechanisms. *Cerebral Cortex, 15*(3), 239–249.

Byrnes, J. P., & Miller, D. C. (2007). The relative importance of predictors of math and science achievement: An opportunity – propensity analysis. *Contemporary Educational Psychology, 32*(4), 599–629.

Byrnes, J. P., & Miller-Cotto, D. (2016). The growth of mathematics and reading skills in segregated and diverse schools: An opportunity-propensity analysis of a national database. *Contemporary Educational Psychology, 46*.

Cai, J., Morris, A., Hohensee, C., Hwang, S., Robison, V., Cirillo, M., . . . Bakker, A. (2020). Addressing the problem of always starting over: Identifying, valuing, and sharing professional knowledge for teaching. *Journal for Research in Mathematics Education, 51*(2), 130–139.

Cai, J., & Wang, T. (2006). U.S. and Chinese teachers' conceptions and constructions of representations: A case of teaching ratio concept. *International Journal of Science and Mathematics Education, 4*(1), 145–186.

Charalambous, C. Y., Hill, H. C., Chin, M. J., & McGinn, D. (2020). Mathematical content knowledge and knowledge for teaching: Exploring their distinguishability and contribution to student learning. *Journal of Mathematics Teacher Education, 23*(6), 579–613.

Charalambous, C. Y., & Praetorius, A.-K. (2018). Studying mathematics instruction through different lenses: Setting the ground for understanding instructional quality more comprehensively. *ZDM, 50*(3), 355–366.

Clark, I., & Dumas, G. (2016). The regulation of task performance: A trans-disciplinary review [Review]. *Frontiers in Psychology, 6*(1862).

Clarke, D., Keitel, C., & Shimizu, Y. (Eds.). (2006). *Mathematics classrooms in twelve countries: The insider's perspective*. Rotterdam: Sense Publishers.

Cohen, D. K., Raudenbush, S. W., & Ball, D. L. (2003). Resources, instruction, and research. *Educational Evaluation and Policy Analysis, 25*(2), 119–142.

Cohen, L., Manion, L., & Morrison, K. (2018). *Research methods in education* (8th ed.). London: Routledge.

Cooke, R. (2013). *The history of mathematics: A brief course.* New York, NY: John Wiley & Sons.

Copur-Gencturk, Y., & Thacker, I. (2021). A comparison of perceived and observed learning from professional development: Relationships among self-reports, direct assessments, and teacher characteristics. *Journal of Teacher Education, 72*(2), 138–151.

Cravens, X. C., & Hunter, S. B. (2021). Assessing the impact of collaborative inquiry on teacher performance and effectiveness. *School Effectiveness and School Improvement, 32*(4), 564–606.

Cravens, X. C., & Wang, J. (2017). Learning from the masters: Shanghai's teacher-expertise infusion system. *International Journal for Lesson and Learning Studies, 6*(4), 306–320.

Croll, P. (1996). Teacher-pupil interaction in the classroom. In P. Croll & N. Hastings (Eds.), *Effective primary teaching* (pp. 14–28). London: David Fulton.

Csikszentmihalyi, M. (1990). *Flow: The psychology of optimal experience.* New York, NY: Harper & Row.

Cumming, M. M., Bettini, E., Pham, A. V., & Park, J. (2020). School-, classroom-, and dyadic-level experiences: A literature review of their relationship with students' executive functioning development. *Review of Educational Research, 90*(1), 47–94.

Daniel, G. R., Wang, C., & Berthelsen, D. (2016). Early school-based parent involvement, children's self-regulated learning and academic achievement: An Australian longitudinal study. *Early Childhood Research Quarterly, 36*, 168–177.

Desoete, A., Baten, E., Vercaemst, V., De Busschere, A., Baudonck, M., & Vanhaeke, J. (2019). Metacognition and motivation as predictors for mathematics performance of Belgian elementary school children. *ZDM, 51*(4), 667–677.

Desoete, A., Roeyers, H., & Buysse, A. (2001). Metacognition and mathematical problem solving in grade 3. *Journal of Learning Disabilities, 34*(5), 435–447.

DeWolf, M., Bassok, M., & Holyoak, K. (2015). From rational numbers to algebra: Separable contributions of decimal magnitude and relational understanding of fractions. *Journal of Experimental Child Psychology, 133*, 72–84.

Di Martino, P., & Zan, R. (2010). 'Me and maths': Towards a definition of attitude grounded on students' narratives. *Journal of Mathematics Teacher Education, 13*(1), 27–48.

DfE UK. (2013). *National curriculum in England: Mathematics programmes of study (Key Stages 1 & 2).* London: Department for Education, UK.

Doğan, S., & Adams, A. (2018). Effect of professional learning communities on teachers and students: Reporting updated results and raising questions about research design. *School Effectiveness and School Improvement, 29*(4), 634–659.

Dossey, J. A. (2017). Problem solving from a mathematical standpoint. In B. Csapó & J. Funke (Eds.), *The nature of problem solving.* Paris: The OECD Publishing.

Downton, A., & Sullivan, P. (2017). Posing complex problems requiring multiplicative thinking prompts students to use sophisticated strategies and build mathematical connections. *Educational Studies in Mathematics, 95*(3), 303–328.

Empson, S. B., Levi, L., & Carpenter, T. P. (2011). The algebraic nature of fractions: Developing relational thinking in elementary school. In J. Cai & E. Knuth (Eds.), *Early algebraization: A global dialogue from multiple perspectives* (pp. 409–428). Heidelberg: Springer.

Fan, L., Miao, Z., & Mok, A. C. I. (2015). How chinese teachers teach mathematics and pursue professional development: Perspectives from contemporary international research. In L. Fan, N.-Y. Wong, J. Cai, & S. Li (Eds.), *How Chinese teach mathematics* (pp. 43–70). Singapore: World Scientific.

Fan, L., Zhu, Y., & Tang, C. (2015). What makes a master teacher? A study of thirty-one mathematics master teachers in chinese mainland. In L. Fan, N.-Y. Wong, J. Cai, & S. Li (Eds.), *How Chinese teach mathematics* (pp. 493–528). Singapore: World Scientific.

Flavell, J. H. (1976). Metacognitive aspects of problem solving. In L. B. Resnick (Ed.), *The nature of intelligence* (pp. 231–235). Hillsdale, NJ: Lawrence Erlbaum Associates.

Flavell, J. H. (1979). Metacognition and cognitive monitoring: A new area of cognitive – developmental inquiry. *American Psychologist, 34*(10), 906–911.

Fleming, S. M., Huijgen, J., & Dolan, R. J. (2012). Prefrontal contributions to metacognition in perceptual decision making. *The Journal of Neuroscience, 32*(18), 6117–6125.

Freudenthal, H. (1983). *Didactical phenomenology of mathematical structures.* Dordrecht: D. Reidel.

Frost, L. A., Hyde, J. S., & Fennema, E. (1994). Gender, mathematics performance, and mathematics-related attitudes and affect: A meta-analytic synthesis. *International Journal of Educational Research, 21*(4), 373–385.

García-García, J., & Dolores-Flores, C. (2018). Intra-mathematical connections made by high school students in performing Calculus tasks. *International Journal of Mathematical Education in Science and Technology, 49*(2), 227–252.

Garon-Carrier, G., Boivin, M., Guay, F., Kovas, Y., Dionne, G., Lemelin, J. P., . . . Tremblay, R. E. (2016). Intrinsic motivation and achievement in mathematics in elementary school: A longitudinal investigation of their association. *Child Development, 87*(1), 165–175.

Garrison, J. (2014). Dissociable neural networks supporting metacognition for memory and perception. *The Journal of Neuroscience, 34*(8), 2765–2767.

Geiger, V., Goos, M., & Forgasz, H. (2015). A rich interpretation of numeracy for the 21st century: A survey of the state of the field. *ZDM, 47*(4), 531–548.

Gladwell, M. (2008). *Outliers: The story of success.* New York, NY: Little, Brown and Company.

Glaser, B. G., & Strauss, A. L. (1967). *The discovery of grounded theory: Strategies for qualitative research.* Chicago, IL: Aldine.

Good, T. L., & Grouws, D. A. (1977). Teaching effects: A process-product study in fourth grade mathematics classrooms. *Journal of Teacher Education, 28*(3), 49–54.

Good, T. L., & Grouws, D. A. (1979). The Missouri mathematics effectiveness project: An experimental study in fourth-grade classrooms. *Journal of Educational Psychology, 71*(3), 355–362.

Grossman, P., & McDonald, M. (2008). Back to the future: Directions for research in teaching and teacher education. *American Educational Research Journal, 45*(1), 184–205.

Gu, L., Huang, R., & Marton, F. (2004). Teaching with variation: A Chinese way of promoting effective mathematics learning. In L. Fan, N. -Y. Wong, J. Cai, & S. Li (Eds.), *How Chinese learn mathematics: Perspectives from insiders.* Singapore: World Scientific Publishing Co.

Hammersley, M. (1993). On the teacher as researcher. *Educational Action Research, 1*(3), 425–445.

Han, X., & Huang, R. (2019). Developing teachers' expertise in mathematics instruction as deliberate practice through Chinese lesson study. In R. Huang, A. Takahashi, & J. P.

da Ponte (Eds.), *Theory and practice of lesson study in mathematics: An international perspective* (pp. 59–81). Cham: Springer.

Han, X., & Paine, L. (2010). Teaching mathematics as deliberate practice through public lessons. *The Elementary School Journal, 110*(4), 519–541.

Hannula, M. S. (2002). Attitude towards mathematics: Emotions, expectations and values. *Educational Studies in Mathematics, 49*(1), 25–46.

Hart, K., Brown, M., Kerslake, D., Küchemann, D. E., & Ruddock, G. (1985a). *Chelsea diagnostic mathematics tests*. Windsor: NFER-Nelson.

Hart, K., Brown, M., Kerslake, D., Küchemann, D. E., & Ruddock, G. (1985b). *Chelsea diagnostic mathematics tests: Teacher's guide*. Windsor: NFER-Nelson.

Hart, K., Brown, M., Küchemann, D. E., Kerslake, D., Ruddock, G., & McCartney, M. (1981). *Children's understanding of mathematics: 11–16*. London: John Murray.

Hart, L. E. (1989). Describing the affective domain: Saying what we mean. In D. B. McLeod & V. M. Adams (Eds.), *Affect and mathematical problem solving: A new perspective* (pp. 37–48). New York, NY: Springer-Verlag.

Hiebert, J., Gallimore, R., Garnier, H., Givvin, K. B., Hollingsworth, H., Jacobs, J., . . . Stigler, J. (2003). *Teaching mathematics in seven countries: Results from the TIMSS 1999 video study*. Washington, DC: National Center for Educational Statistics.

Hill, H. C., & Ball, D. L. (2004). Learning mathematics for teaching: Results from California's mathematics professional development institutes. *Journal for Research in Mathematics Education, 35*(5), 330–351.

Hill, H. C., Ball, D. L., & Schilling, S. G. (2008). Unpacking pedagogical content knowledge: Conceptualizing and measuring teachers' topic-specific knowledge of students. *Journal for Research in Mathematics Education, 39*(4), 372–400.

Hill, H. C., & Chin, M. (2018). Connections between teachers' knowledge of students, instruction, and achievement outcomes. *American Educational Research Journal, 55*(5), 1076–1112.

Hodgen, J., Foster, C., Marks, R., & Brown, M. (2018). *Improving mathematics in key stages two and three: Evidence review*. London: Education Endowment Foundation.

Hodgen, J., Küchemann, D., Brown, M., & Coe, R. (2010). Multiplicative reasoning, ratio and decimals: A 30-year comparison of lower secondary students' understandings. In M. M. F. Pinto & T. F. Kawasaki (Eds.), *Proceedings of the 34th conference of the international group for the psychology of mathematics education* (Vol. 3, pp. 89–96). Belo Horizonte, Brazil: PME.

Hodgkin, L. (2005). *A history of mathematics: From Mesopotamia to modernity*. Oxford: Oxford University Press.

Houston, K. (2009). *How to think like a mathematician: A companion to undergraduate mathematics*. Cambridge: Cambridge University Press.

Howe, C., Luthman, S., Ruthven, K., Mercer, N., Hofmann, R., Ilie, S., & Guardia, P. (2015). Rational number and proportional reasoning in early secondary school: Towards principled improvement in mathematics. *Research in Mathematics Education, 17*(1), 38–56.

Hu, L., & Bentler, P. M. (1999). Cutoff criteria for fit indexes in covariance structure analysis: Conventional criteria versus new alternatives. *Structural Equation Modeling: A Multidisciplinary Journal, 6*(1), 1–55.

Huang, R., & Bao, J. (2006). Towards a model for teacher professional development in China: Introducing Keli. *Journal of Mathematics Teacher Education, 9*(3), 279–298.

Huang, R., & Leung, F. K. S. (2005). Deconstructing teacher-centeredness and student-centeredness dichotomy: A case study of a Shanghai mathematics lesson. *The Mathematics Educator, 15*(2), 35–41.

Huang, R., & Li, Y. (2009). Pursuing excellence in mathematics classroom instruction through exemplary lesson development in China: A case study. *ZDM, 41*(3), 297–309.

Huang, R., & Li, Y. (Eds.). (2017). *Teaching and learning mathematics through variation: Confucian heritage meets Western theories*. Rotterdam: Sense Publishers.

Huang, R., Prince, K. M., Barlow, A. T., & Schmidt, T. (2017). Improving mathematics teaching as deliberate practice through Chinese lesson study. *The Mathematics Educator, 26*(1), 32–55.

Huang, R., Ye, L., & Prince, K. (2017). Professional development of secondary matehamtics teachers in mainland China. In B. Kaur, O. N. Kwon, & Y. H. Leong (Eds.), *Professional development of mathematics teachers: An Asian perspective* (pp. 17–31). Singapore: Springer Nature Singapore.

Hurst, M., & Cordes, S. (2018). A systematic investigation of the link between rational number processing and algebra ability. *British Journal of Psychology, 109*(1), 99–117.

IEA. (2014). *TIMSS 2015 student questionnaire (grade 4)*. Chestnut Hill, MA: TIMSS & PIRLS International Study Center, Boston College.

Jacobs, J. K., Garnier, H., Gallimore, R., Hollingsworth, H., Givvin, K. B., Rust, K., . . . Manaster, A. (2003). *TIMSS 1999 video study technical report*. Washington, DC: National Centre for Education Statistics, US Department of Education.

Jensen, B., Sonnemann, J., Roberts-Hull, K., & Hunter, A. (2016). *Beyond PD: Teacher professional learning in high-performing systems*. Washington, DC: National Center on Education and the Economy.

Jiang, Z., Mok, I. A. C., & Li, J. (2021). Chinese students' hierarchical understanding of part-whole and measure subconstructs. *International Journal of Science and Mathematics Education, 19*, 1441–1461.

Kenny, D. A., Kaniskan, B., & McCoach, D. B. (2015). The performance of RMSEA in models with small degrees of freedom. *Sociological Methods & Research, 44*(3), 486–507.

Kiefer, T., Robitzsch, A., & Wu, M. (2021). TAM (Test analysis modules): An R package. *TAM Tutorials*. http://www.edmeasurementsurveys.com/TAM/Tutorials/.

Kim, Y. R., Park, M. S., Moore, T. J., & Varma, S. (2013). Multiple levels of metacognition and their elicitation through complex problem-solving tasks. *The Journal of Mathematical Behavior, 32*(3), 377–396.

King, R., & McInerney, D. (2016). Do goals lead to outcomes or can it be the other way around? Causal ordering of mastery goals, metacognitive strategies, and achievement. *British Journal of Educational Psychology, 86*(3), 296–312.

Kiwanuka, H. N., Van Damme, J., Van Den Noortgate, W., Anumendem, D. N., Vanlaar, G., Reynolds, C., & Namusisi, S. (2017). How do student and classroom characteristics affect attitude toward mathematics? A multivariate multilevel analysis. *School Effectiveness and School Improvement, 28*(1), 1–21.

Kline, R. B. (2005). *Principles and practice of structural equation modeling* (2nd ed.). New York, NY: Guilford Press.

Köğce, D., Yıldız, C., Aydın, M., & Altındağ, R. (2009). Examining elementary school students' attitudes towards mathematics in terms of some variables. *Procedia – Social and Behavioral Sciences, 1*(1), 291–295.

Kraft, M. A., & Hill, H. C. (2020). Developing ambitious mathematics instruction through web-based coaching: A randomized field trial. *American Educational Research Journal, 57*(6), 2378–2414.

Krantz, S. G. (2010). *An episodic history of mathematics: Mathematical culture through problem solving*. Washington, DC: The Mathematical Association of America.

Kuzle, A. (2018). Assessing metacognition of grade 2 and grade 4 students using an adaptation of multi-method interview approach during mathematics problem-solving. *Mathematics Education Research Journal*, *30*(2), 185–207.

Kyriakides, L., Creemers, B. P. M., Panayiotou, A., Vanlaar, G., Pfeifer, M., Cankar, G., & McMahon, L. (2014). Using student ratings to measure quality of teaching in six European countries. *European Journal of Teacher Education*, *37*(2), 125–143.

Lamon, S. J. (1993). Ratio and proportion: Connecting content and children's thinking. *Journal for Research in Mathematics Education*, *24*(1), 41–61.

Lamon, S. J. (1995). Ratio and proportion: Elementary didactical phenomenology. In J. T. Sowder & B. P. Schappelle (Eds.), *Providing a foundation for teaching mathematics in the middle grades* (pp. 167–198). Albany, NY: State University of New York Press.

Lamon, S. J. (2007). Rational numbers and proportional reasoning: Toward a theoretical framework for research. In F. K. Lester (Ed.), *Second handbook of research on mathematics teaching and learning: A project of the national council of teachers of mathematics* (pp. 629–667). Washington, DC: National Council of Teachers of Mathematics.

Lampert, M. (2001). *Teaching problems and the problems of teaching*. New Haven, CT: Yale University Press.

Landis, J. R., & Koch, G. G. (1977). The measurement of observer agreement for categorical data. *Biometrics*, *33*(1), 159–174.

Learning Mathematics for Teaching, P. (2011). Measuring the mathematical quality of instruction. *Journal of Mathematics Teacher Education*, *14*(1), 25–47.

Leinhardt, G. (1989). Math lessons: A contrast of novice and expert competence. *Journal for Research in Mathematics Education*, *20*(1), 52–75.

Lester, F. K., Garofalo, J., & Kroll, D. L. (1989). Self-confidence, interest, beliefs, and metacognition: Key influences on problem-solving behaviour. In D. B. Mcleod & V. M. Adams (Eds.), *Affect and mathematical problem solving: A new perspective* (pp. 75–88). New York, NY: Springer-Verlag.

Leuders, T., Philipp, K., & Leuders, J. (Eds.). (2018). *Diagnostic competence of mathematics teachers: Unpacking a complex construct in teacher education and teacher practice*. Cham: Springer.

Leung, F. K. S. (1992). *A comparison of the intended mathematics curriculum in China, Hong Kong and England and the implementation in Beijing, Hong Kong and London* [Doctoral dissertation, University of London].

Leung, F. K. S. (2005). Some characteristics of East Asian mathematics classrooms based on data from the TIMSS 1999 video study. *Educational Studies in Mathematics*, *60*(2), 199–215.

Li, J. (2004). Thorough understanding of the textbook: A significant feature of Chinese teacher manuals. In L. Fan, N. Y. Wong, J. Cai, & S. Li (Eds.), *How Chinese learn mathematics: Perspectives from insiders* (pp. 262–281). Singapore: World Scientific Publishing.

Li, J., Hu, Y., & Fan, W. (2016). Family background, shadow education and student achievement: An empirical study based on the Wisconsin model [In Chinese]. *China Economics of Education Review*, *1*(01), 70–89.

Li, Y. (2002). Knowing, understanding and exploring the content and formation of curriculum materials: A Chinese approach to empower prospective elementary school teachers pedagogically. *International Journal of Educational Research*, *37*(2), 179–193.

Li, Y., Tang, C., & Gong, Z. (2011). Improving teacher expertise through master teacher work stations: A case study. *ZDM*, *43*(6), 763–776.

Liu, R.-D., Ding, Y., Zong, M., & Zhang, D. (2014). Concept development of decimals in Chinese elementary students: A conceptual change approach. *School Science and Mathematics*, *114*(7), 326–338.

Luyten, H. (2003). The size of school effects compared to teacher effects: An overview of the research literature. *School Effectiveness and School Improvement*, *14*(1), 31–51.

Ma, L. (1999). *Knowing and teaching elementary mathematics: Teachers' understanding of fundamental mathematics in China and the United States*. Mahwah, NJ: Lawrence Erlbaum.

Ma, X., & Kishor, N. (1997). Assessing the relationship between attitude toward mathematics and achievement in mathematics: A Meta-analysis. *Journal for Research in Mathematics Education*, *28*(1), 26–47.

Ma, X., & Xu, J. (2004). The causal ordering of mathematics anxiety and mathematics achievement: A longitudinal panel analysis. *Journal of Adolescence*, *27*(2), 165–179.

McDonald, R. P., & Ho, M.-H. R. (2002). Principles and practice in reporting structural equation analyses. *Psychological Methods*, *7*(1), 64.

Merzbach, U. C., & Boyer, C. B. (2011). *A history of mathematics* (3rd ed.). Hoboken, NJ: John Wiley & Sons.

Metallidou, P., & Vlachou, A. (2007). Motivational beliefs, cognitive engagement, and achievement in language and mathematics in elementary school children. *International Journal of Psychology*, *42*(1), 2–15.

Miao, Z., Bokhove, C., Reynolds, D., & Charalambous, C. Y. (2022). Rational numbers and proportional reasoning in Chinese primary schools: Patterns, latent classes and reasoning processes. *Asian Journal for Mathematics Education*, *1*(4).

Miao, Z., & Reynolds, D. (2015, November 2). Effectiveness of mathematics teaching: The truth about China and England. *The BERA Blog*. www.bera.ac.uk/blog/effectiveness-of-mathematics-teaching-the-truth-about-china-and-england.

Miao, Z., & Reynolds, D. (2018). *The effectiveness of mathematics teaching in primary schools: Lessons from England and China*. London: Routledge.

Miao, Z., Reynolds, D., & Bokhove, C. (2021, July 12–18). *First voyage of the integrated paradigm: The case of an international study on effective mathematics teaching* [Paper presentation]. The 14th International Congress on Mathematical Education (ICME-14), Shanghai, China.

Miao, Z., Reynolds, D., Harris, A., & Jones, M. (2015). Comparing performance: A cross-national investigation into the teaching of mathematics in primary classrooms in England and China. *Asia Pacific Journal of Education*, *35*(3), 392–403.

Minato, S., & Kamada, T. (1996). Results of research studies on causal predominance between achievement and attitude in junior high school mathematics of Japan. *Journal for Research in Mathematics Education*, *27*(1), 96–99.

MoE China. (2001). *Mathematics curriculum standards for nine-year compulsory education (Experimental version)*. Beijing: Beijing Normal University [in Chinese].

MoE China. (2011). *Mathematics curriculum standards for nine-year compulsory education*. Beijing: Beijing Normal University [in Chinese].

MoE China. (2022). *Mathematics curriculum standards for nine-year compulsory education*. Beijing: Beijing Normal University [in Chinese].

MoE Singapore. (2012). *Mathematics syllabus: Primary One to Six*. Singapore: Curriculum Planning and Development Division of MoE.

Mohamed, L., & Waheed, H. (2011). Secondary students' attitude towards mathematics in a selected school of Maldives. *International Journal of Humanities and Social Science*, *1*(15), 277–281.

Mohd, N., Mahmood, T., & Ismail, M. N. (2011). Factors that influence students in mathematics achievement. *International Journal of Academic Research, 3*(3), 49–54.

Mok, I. A. C. (2006). Shedding light on the East Asian learner paradox: Reconstructing student-centredness in a Shanghai classroom. *Asia Pacific Journal of Education, 26*(2), 131–142.

Mortimore, P., Sammons, P., Stoll, L., Lewis, D., & Ecob, R. (1988). *School matters*. Berkeley, CA: The University of California Press.

Moseley, B. J., Okamoto, Y., & Ishida, J. (2007). Comparing US and Japanese elementary school teachers' facility for linking rational number representations. *International Journal of Science and Mathematics Education, 5*(1), 165–185.

Muijs, D., & Bokhove, C. (2020). *Metacognition and self-regulation: Evidence review*. Education Endowment Foundation. The report is available from: https://educationendowmentfoundation.org.uk/education-evidence/evidence-reviews/metacognition-and-self-regulation.

Muijs, D., Kyriakides, L., van der Werf, G., Creemers, B. P. M., Timperley, H., & Earl, L. (2014). State of the art – teacher effectiveness and professional learning. *School Effectiveness and School Improvement, 25*(2), 231–256.

Muijs, D., & Reynolds, D. (2000). School effectiveness and teacher effectiveness in mathematics: Some preliminary findings from the evaluation of the mathematics enhancement programme (primary). *School Effectiveness and School Improvement, 11*(3), 273–303.

Muijs, D., & Reynolds, D. (2003). Student background and teacher effects on achievement and attainment in mathematics: A longitudinal study. *Educational Research and Evaluation: An International Journal on Theory and Practice, 9*(3), 289–314.

Muijs, D., Reynolds, D., Sammons, P., Kyriakides, L., Creemers, B. P. M., & Teddlie, C. (2018). Assessing individual lessons using a generic teacher observation instrument: How useful is the international system for teacher observation and feedback (ISTOF)? *ZDM, 50*(3), 395–406.

Mullis, I. V. S., Martin, M. O., Foy, P., & Hooper, M. (2016). *TIMSS 2015 international results in mathematics*. Chestnut Hill, MA: Boston College and IEA.

Muthukrishna, M., Bell, A. V., Henrich, J., Curtin, C. M., Gedranovich, A., McInerney, J., & Thue, B. (2020). Beyond western, educated, industrial, rich, and democratic (WEIRD) psychology: Measuring and mapping scales of cultural and psychological distance. *Psychological Science, 31*(6), 678–701.

National Bureau of Statistics of China. (2019). *Gross domestic product of major cities in 2018*. https://data.stats.gov.cn/english/easyquery.htm?cn=E0105.

Neel, C. G.-O., & Fuligni, A. (2013). A longitudinal study of school belonging and academic motivation across high school. *Child Development, 84*(2), 678–692.

NGA & CCSSO. (2010). *Common core state standards for mathematics*. Washington, DC: Authors.

Ning, H. K. (2016). Examining heterogeneity in student metacognition: A factor mixture analysis. *Learning and Individual Differences, 49*, 373–377.

Nistal, A., Dooren, W., Clarebout, G., Elen, J., & Verschaffel, L. (2009). Conceptualising, investigating and stimulating representational flexibility in mathematical problem solving and learning: A critical review. *ZDM, 41*, 627–636.

NMAP. (2008). *Foundations for success: The final report of the national mathematics advisory panel*. Washington, DC: U.S. Department of Education.

Noss, R., Healy, L., & Hoyles, C. (1997). The construction of mathematical meanings: Connecting the visual with the symbolic. *Educational Studies in Mathematics, 33*(2), 203–233.

Nuffield Foundation. (1994). *History of mathematics*. Singapore: Longman.

OECD. (2013). *Teaching and learning international survey (TALIS) 2013: Teacher questionnaire*. Paris: OECD Publishing.

OECD. (2018). *OECD future of education and skills 2030: Conceptual learning framework*. Paris: The OECD Publishing.

OECD. (2019). *PISA 2018 results: What students know and can do* (Vol. 1). Paris: The OECD Publishing.

OECD. (2020). *Global teaching insights: A video study of teaching*. OECD Publishing.

Ohtani, K., & Hisasaka, T. (2018). Beyond intelligence: A meta-analytic review of the relationship among metacognition, intelligence, and academic performance. *Metacognition and Learning, 13*(2), 179–212.

Paine, L., & Ma, L. (1993). Teachers working together: A dialogue on organizational and cultural perspectives of Chinese teachers. *International Journal of Educational Research, 19*(8), 675–718.

Pappas, S., Ginsburg, H. P., & Jiang, M. (2003). SES differences in young children's metacognition in the context of mathematical problem solving. *Cognitive Development, 18*(3), 431–450.

Peck, F. A. (2020). Beyond rise over run: A learning trajectory for slope. *Journal for Research in Mathematics Education, 51*(4), 433.

PEP. (2014a). *Mathematics (5B)*. Beijing: People's Education Press [in Chinese].

PEP. (2014b). *Mathematics (6B)*. Beijing: People's Education Press [in Chinese].

PhEP. (2014a). *Mathematics (3B)*. Nanjing: Phoenix Education Publishing [in Chinese].

PhEP. (2014b). *Mathematics (4B)*. Nanjing: Phoenix Education Publishing [in Chinese].

PhEP. (2014c). *Mathematics (5B)*. Nanjing: Phoenix Education Publishing [in Chinese].

PhEP. (2014d). *Mathematics (6B)*. Nanjing: Phoenix Education Publishing [in Chinese].

Pressley, M., & Harris, K. R. (2006). Cognitive strategy instruction: From basic research to classroom instruction. In P. A. Alexander & P. Winne (Eds.), *Handbook of educational psychology* (2nd ed., pp. 265–286). Mahwah, NJ: Erlbaum.

R Core Team. (2021). *R: A language and environment for statistical computing*. R Foundation for Statistical Computing. www.R-project.org/.

Reynolds, D., Creemers, B. P. M., Stringfield, S., Teddlie, C., & Schaffer, G. (Eds.). (2002). *World class schools: International perspectives on school effectiveness*. London: Routledge.

Rittle-Johnson, B., Schneider, M., & Star, J. R. (2015). Not a one-way street: Bidirectional relations between procedural and conceptual knowledge of mathematics. *Educational Psychology Review, 27*(4), 587–597.

Rosseel, Y. (2012). Lavaan: An R package for structural equation modeling and more. Version 0.5–12 (BETA). *Journal of Statistical Software, 48*(2), 1–36.

RStudio Team. (2021). *RStudio: Integrated development for R*. RStudio, PBC. www.rstudio.com.

Scheerens, J., Vermeulen, C. J. A. J., & Pelgrum, W. J. (1989). Generalizability of instructional and school effectiveness indicators across nations. *International Journal of Educational Research, 13*(7), 789–799.

Scherer, R., & Gustafsson, J.-E. (2015). Student assessment of teaching as a source of information about aspects of teaching quality in multiple subject domains: An application of multilevel bifactor structural equation modeling. *Frontiers in Psychology, 6*, 1–15.

Scherer, R., & Nilsen, T. (2016). The relations among school climate, instructional quality, and achievement motivation in mathematics. In T. Nilsen & J.-E. Gustafsson (Eds.),

Teacher quality, instructional quality and student outcomes: Relationships across countries, cohorts and time (pp. 51–80). Cham: Springer International Publishing.

Schneider, W., & Artelt, C. (2010). Metacognition and mathematics education. *ZDM, 42*(2), 149–161.

Schnotz, W., & Bannert, M. (2003). Construction and interference in learning from multiple representation. *Learning and Instruction, 13*, 141–156.

Schoenfeld, A. H. (1989). Explorations of students' mathematical beliefs and behavior. *Journal for Research in Mathematics Education, 20*(4), 338–355.

Schoenfeld, A. H. (1992). Learning to think mathematically: Problem solving, metacognition, and sense-making in mathematics. In D. Grouws (Ed.), *Handbook for research on mathematics teaching and learning* (pp. 334–370). New York, NY: Macmillan.

Schön, D. A. (1983). *The reflective practitioner: How professionals think in action.* New York, NY: Basic Books.

Shulman, L. (1986). Those who understand: Knowledge growth in teaching. *Educational Researcher, 15*(2), 4–14.

Sidney, P. G., & Alibali, M. W. (2015). Making connections in math: Activating a prior knowledge analogue matters for learning. *Journal of Cognition and Development, 16*(1), 160–185.

Siegler, R. S., Duncan, G. J., Davis-Kean, P. E., Duckworth, K., Claessens, A., Engel, M., . . . Chen, M. (2012). Early predictors of high school mathematics achievement. *Psychological Science, 23*(7), 691–697.

Sims, S., & Fletcher-Wood, H. (2021). Identifying the characteristics of effective teacher professional development: A critical review. *School Effectiveness and School Improvement, 32*(1), 47–63.

Slaten, C. D., Ferguson, J. K., Allen, K.-A., Brodrick, D.-V., & Waters, L. (2016). School belonging: A review of the history, current trends, and future directions. *The Educational and Developmental Psychologist, 33*(1), 1–15.

Sperling, R. A., Howard, B. C., Miller, L. A., & Murphy, C. (2002). Measures of children's knowledge and regulation of cognition. *Contemporary Educational Psychology, 27*(1), 51–79.

Stevenson, H. W., & Stigler, J. W. (1992). *The learning gap: Why our schools are failing, and what we can learn from Japanese and Chinese education.* New York: Summit Books.

Stigler, J. W., & Hiebert, J. (1999). *The teaching gap: Best ideas from the world's teachers for improving education in the classroom.* New York, NY: The Free Press.

Teddlie, C., Creemers, B. P. M., Kyriakides, L., Muijs, D., & Fen, Y. (2006). The international system for teacher observation and feedback: Evolution of an international study of teacher effectiveness constructs. *Educational Research and Evaluation, 12*(6), 561–582.

Teffer, K., & Semendeferi, K. (2012). Human prefrontal cortex: Evolution, development, and pathology. *Progress in Brain Research, 195*, 191–218.

Thapa, A., Cohen, J., Guffey, S., & Higgins-D'Alessandro, A. (2013). A review of school climate research. *Review of Educational Research, 83*(3), 357–385.

Valverde, G. A., Bianchi, L. J., Wolfe, R. G., Schmidt, W. H., & Houang, R. T. (2002). *According to the book: Using TIMSS to investigate the translation of policy into practice through the world of textbooks.* Dordrecht: Kluwer Academic Publishers.

Veenman, M. V., Elshout, J. J., & Meijer, J. (1997). The generality vs domain-specificity of metacognitive skills in novice learning across domains. *Learning and Instruction, 7*(2), 187–209.

Veenman, M. V., Van Hout-Wolters, B. H., & Afflerbach, P. (2006). Metacognition and learning: Conceptual and methodological considerations. *Metacognition and Learning, 1*(1), 3–14.

Wang, A. H., Shen, F., & Byrnes, J. P. (2013). Does the opportunity – propensity framework predict the early mathematics skills of low-income pre-kindergarten children? *Contemporary Educational Psychology, 38*(3), 259–270.

Wong, J. L. N. (2010). Searching for good practice in teaching: A comparison of two subject-based professional learning communities in a secondary school in Shanghai. *Compare: A Journal of Comparative and International Education, 40*(5), 623–639.

Wu, M., Tam, H. P., & Jen, T.-H. (2016). *Educational measurement for applied researchers: Theory into practice*. Singapore: Springer Nature Singapore.

Xie, Y., Zhang, X., Tu, P., & Ren, Q. (2017). *China family panel studies* (3rd ed.). Beijing: Peiking University.

Yoon, K. S., Duncan, T., Lee, S. W.-Y., Scarloss, B., & Shapley, K. L. (2007). *Reviewing the evidence on how teacher professional development affects student achievement*. Washington, DC: Regional Educational Laboratory Southwest (NJ1).

Zhang, M., Walker, A. D., & Qian, H. (2021). Master teachers as system leaders of professional learning: Master teacher studios in China. *International Journal of Educational Management, 35*(7), 1333–1346.

Zhao, N., Teng, X., Li, W., Li, Y., Wang, S., Wen, H., & Yi, M. (2019). A path model for metacognition and its relation to problem-solving strategies and achievement for different tasks. *ZDM, 51*(4), 641–653.

Zheng, X., Yin, H., & Liu, Y. (2020). Are professional learning communities beneficial for teachers? A multilevel analysis of teacher self-efficacy and commitment in China. *School Effectiveness and School Improvement*, 1–21.

INDEX

Note: Page numbers in *italic* indicate a figure and page numbers in **bold** indicate a table on the corresponding page.